# Ecosystem Geography
## From Ecoregions to Sites

Second Edition

**Frontispiece.** The semiarid mountains of the Southern Rocky Mountains, near Crested Butte, Colorado, contain numerous examples of the landform influences on ecosystem patterns. *Slate River*, oil on canvas, plein air, by Shaun Horne © 2004, reproduced with permission.

Robert G. Bailey

# Ecosystem Geography

## From Ecoregions to Sites

Second Edition

With a Foreword by Jack Ward Thomas, Chief, USDA Forest Service

Illustrations by Lev Ropes and Nancy Maysmith

With 142 Illustrations in 154 Parts

 Springer

Robert G. Bailey
Rocky Mountain Research Station
USDA Forest Service
240 W Prospect Road
Fort Collins, CO 80526
USA
rgbailey@fs.fed.us

Additional material to this book can be downloaded from http://extra.springer.com

ISBN 978-0-387-89515-4    (hardcover)          e-ISBN 978-0-387-89516-1
ISBN 978-1-4419-0391-4    (softcover)
DOI 10.1007/978-0-387-89516-1
Springer New York Dordrecht Heidelberg London

Library of Congress Control Number: 2009928500

*Cover illustration (clockwise from top left):* Ecoregion divisions map (level 2 of the ecoregion hierarchy) of the
world (Source: Microsoft Virtual Globe ©1995–1998, Microsoft Corporation; map reprinted with permission
from Microsoft Corporation) – Landscape mosaic consisting of spruce forests and glacially scoured lakes,
warm continental division, Minnesota (Photograph by Jack Boucher, National Park Service) – Montane forest
site, subtropical steppe mountains, Arizona (Photograph by Robert G. Bailey) – Semidesert site, subtropical
steppe division, Arizona (Photograph by Robert G. Bailey) – Desert site, subtropical desert division, Arizona
(Photograph by Robert G. Bailey)

Printed on acid-free paper

Springer is part of Springer Science+Business Media (www.springer.com)

# Foreword

Land management is presently undergoing enormous change: away from managing single resources to managing ecosystems. From forest to tundra, to desert, to steppe, the world's ecosystems vary vastly. To manage them effectively we need to understand their geographic distribution better. We need to do this at various levels of detail because ecosystems exist at multiple scales in a hierarchy, from regional to local.

Maps are needed to display ecosystem distribution and hierarchy. Until now, information on defining ecosystem boundaries has been scarce. This book is the first to clarify and systematize the underlying principles for their mapping. It presents a synthesis of the knowledge in this field and provides a guide to its use.

I recommend this book to all who are involved in the study and management of ecosystems.

Chief, USDA Forest Service                                  Jack Ward Thomas

# Preface to the Second Edition

This book outlines a system that organizes the Earth into a hierarchy of increasingly finer-scale ecosystems that can serve as a consistent framework for ecological analysis and management. The system consists of a three-level hierarchy of nested ecosystem units and their associated mapping criteria. Delineation of units involves identifying the environmental factors controlling the spatial geography of ecosystems at various levels and establishing boundaries where these factors change significantly. Macroscale units (*ecoregions*) are climatically controlled and delineated as Köppen–Trewartha climate zones. Nested within these are *landscape mosaics*, the mesoscale units, controlled by landform and delineated by Hammond's landform regions. At the microscale are individual *sites* controlled by topographically determined topoclimate and soil moisture regimes.

The first edition of this work (1996) was written at a time when few published materials on ecosystem geography were available, and none of these had systematically elaborated the principles underlying the mapping of ecosystems in a form accessible to advanced students and practitioners. This second edition builds on the strengths of its predecessor, incorporates new information, clarifies concepts presented in the first edition, and contains new sections.

The new sections address how ecoregion boundaries were determined, ecoregion redistribution under climate change, ecosystem processes (such as fire regimes), empirical versus genetic approaches to classification, and human modification to ecosystems (for instance, through the introduction of invasive species).

Once again, I would like to thank many people who have made the completion of this book possible: Nancy Maysmith for re-creating many of the first edition diagrams and drawing several new ones, and to Shaun Horne for the frontispiece; Michael Wilson and Renee O'Brien, Program Manager and Deputy Program Manager, respectively, for Inventory, Monitoring, and Analysis at the Rocky Mountain Research Station, for their support; and Eric Smith of the U.S. Forest Service for his review and suggested improvements in the section on climate change. I appreciate the helpful criticism of several reviewers of the first edition, but I should mention especially Richard Huggett, Hartmut Leser, Randy Rosiere, Robert Smith, Duane Griffin, Kenneth Young, John Fedkiw, Steven Jennings, David Scarnecchia, Fred Smeins, and Melinda Knutson. As always, it has been a pleasure to work with Janet Slobodien at Springer in translating this work to print.

Fort Collins, Colorado                                    Robert G. Bailey
December 2008

# Preface to the First Edition

The management of public land needs a new approach. To fill this need, many public land-management agencies in the United States and abroad are working toward the management of ecosystems rather than the management of individual resources. Historically, the ecosystem has been defined as a small homogeneous area, such as a stand of trees or a meadow. Today there are several reasons for recognizing ecosystems at broader scales. Because of the linkages between systems, a modification of one system may affect the operation of surrounding systems. Furthermore, how a system will respond to management is partially determined by relationships with surrounding systems. Understanding these relationships is important in analyzing cumulative effects, with action at one scale and effects at another. This has created the need to subdivide the land into ecosystems of different size (or scale) based on how geographically related systems are linked. This book explores a new approach: one involving ecosystem geography, the study of the distribution and structure of ecosystems as interacting spatial units at various scales, and the processes that have differentiated them.

The basic concepts about scale and ecosystems are discussed in textbooks on landscape ecology and geography (cf. Isachenko 1973; Leser 1976; Forman and Godron 1986). I have presented a synthesis of these concepts elsewhere (Bailey 1985). In follow-up publications (Bailey 1987, 1988a), I suggested criteria for subdividing land areas into ecosystems and provided a discussion of applications. I also showed how existing information and maps could be used to map ecosystems. The scheme that serves as the framework of this book was first devised as a training program for my course in multiscale ecosystem analysis for the U.S.

Forest Service. This publication updates and expands the knowledge of the subject.

My thanks to David H. Miller and J. Stan Rowe; their work was the intellectual background for this book. I would also like to extend thanks for the inspiration provided by John M. Crowley, who introduced me to the fascination of ecosystem geography.

Lev and Linda Ropes helped me to elaborate and illustrate the ideas that help hold this book together. The maps were made by Jon Havens, whose skill and patience have been invaluable. I am also indebted for some of the drawings to Susan Strawn, who also was an alert critic.

Fort Collins, Colorado                                    Robert G. Bailey
March 1995

# Contents

# CHAPTER 1

# Introduction

Beginning with the Resources Planning Act of 1974 (Public Law 93–378), several pieces of legislation require federal land-management agencies to inventory the renewable resources of the nation. Data from the inventory must accurately describe the current conditions, present and potential production levels, and current and prospective use of the individual resources. Data collected in the inventory provide estimates of such information as volume of timber, pounds of available forage, plant species composition, soil depth, wildlife and fish habitat characteristics, land ownership, and land descriptors, such as slope, aspect, and topography. The information that describes current condition and productive potential of each resource is needed to evaluate alternative management strategies with respect to cost, returns, and changes in production.

Such a large body of information is usable only if arranged systematically. Land classification is the process of arranging or ordering information about land units so we can better understand their similarities and relationships (Bailey et al. 1978). Recognition that classification is meaningful in resource inventory is not new. Decades of research and field operations by a host of practitioners have produced classifications that deal with resources as singular and independent items. What is needed now is a classification that provides a basis for a firm understanding of the relationships and interactions between different resources on the *same unit of land*. Several interdisciplinary committees have been established over the past two decades to find a system for classifying and mapping land units that would satisfy the need for a more integrated ecological approach. These efforts have had only limited success because they have had to deal with several significant problems.

R.G. Bailey, *Ecosystem Geography*, DOI 10.1007/978-0-387-89516-1_1,
© Springer Science+Business Media, LLC 2009

# The Problems

Some renewable resources have been inventoried since the late 1800s. These inventories were designed primarily to assess individual resources for a specific purpose. The quality and quantity of available data vary. Timber was inventoried extensively, whereas other resources, such as wildlife and recreation, received little attention. Increasing demand for all resources requires decisions that cannot be made using the existing classification of land units. Examples of specific problems include the following.

Resource inventories generally have not been coordinated. The overlap among estimates of resource production is impossible to determine. For example, estimates of timber potential and livestock grazing potential are available, but it is difficult to determine from existing data whether these potentials involve the same acreages.

Resource data exist as disconnected bits of descriptive information for the purpose of answering specific functional questions. However, because the management and use of one resource often simultaneously affects other resources, this interaction must be taken into consideration. Existing inventories only give a picture of resource composition; they give no understanding of how resources are integrated and interact on the landscape.

Managers have problems trying to base decisions on disconnected information from several single-resource inventories. This is because land is not managed on an individual-resource basis. It is, or should be, managed as an integrated entity with a full range of biotic and abiotic characteristics.

We need resource data for several levels of planning, ranging from the national to the local level. Many inventories are designed to guide on-the-ground management activities of action agencies. However, even local activities must be based not only on the local ecological conditions but also on how such local conditions fit into a broader context. This is because relationships with adjoining areas partially determine the response of a piece of land to management. Existing inventories are not conducted with reference to a hierarchy of ecological land units and cannot aid in assessing the impact of management practices on adjacent or interrelated land units.

The impact of these problems on the inventory and assessment of resources can be reduced by developing a classification and mapping system that captures the integrated nature of the land's resources. Such a system should also be understandable in relation to surrounding land units in a spatial hierarchy.

Attempts to develop such a system have encountered several difficulties. Integration has been a major problem. How the various physical and biotic components are integrated on a piece of land cannot be determined solely by analysis of its components. Another major problem is formulating a common land unit for the many prospective users. For example, certain land components, such as the status of soil nutrients, must be included for foresters but may be of marginal interest to engineers. The set of characteristics chosen as significant for classifying an ecological unit for one resource use must often be revised to suit another purpose. The result is likely to be a different pattern of units for each activity considered.

This fragmented approach to ecosystem classification is not going to satisfy the need for integrated information about the ecosystem and its resources. The expense alone of collecting separate information on timber, wildlife, recreation, and other resources precludes it. In the United States, we must consider interaction among these separate outputs on the same unit of land to comply with environmental laws and multiple-use mandates. For these reasons, a general multipurpose classification system is needed. This does not mean that special purpose, functional classification (e.g., forest type) of land units will no longer be needed. They will, but they should be done *within the context of the multipurpose system.*

# Where Are We Headed?

The problem is to find a system that classifies land as integrated entities but is still suitable for multipurpose applications. In the United States over the past two decades, work to develop such an integrated classification has involved the ecosystem concept (Schultz 1967; Van Dyne 1969). This, in turn, has become an important part of the ecosystem management process in many federal agencies. (For a discussion of ecosystem perspectives of multiple-use management, see U.S. General Accounting Office 1994 and the series of articles in *Ecological Applications* no. 3 1992). The kinds of ecosystems vary vastly in many ways, including their ability to sustain use impacts. A footprint in a rainforest might disappear after half an hour, but in the Antarctic, it might take 10 years. To manage ecosystems effectively, we need to delineate their boundaries. *Ecological land classification* refers to an integrated approach that divides landscapes into ecosystem units of various sizes.

# The Ecosystem Approach

In simple terms, the ecosystem concept proposes that the earth operates as a series of interrelated systems within which all components are linked, so that a change in any one component may bring about some corresponding change in other components and in the operation of the whole system (Fig. 1.1). An ecosystem approach to land evaluation stresses the interrelationships among components rather than treating each one as a separate characteristic of the landscape. It provides a basis for making predictions about resource interaction (e.g., the effects of timber harvesting on water quality).

J.S. Rowe (1961) defined an ecosystem as "a topographic unit, a volume of land and air plus organic contents extended areally over a particular part of the earth's surface for a certain time." This definition stresses the reality of ecosystems as geographic units of the landscape that include all natural phenomena and that can be identified and surrounded by boundaries.

# Classification of Land as Ecosystems

Ecologists and geographers have proposed and classified land as systems for resource management ever since Arthur Tansley (1935) coined the term *ecosystem*. However, the concept of land as an ecosystem is much older. The ancient Greeks recognized such a concept. In the 18th century, Baron von Humboldt provided an outline of latitudinal zonality and high-altitude zonality of the plant and animal world in relation to climate (Berghaus 1845). The significant work of Vasily Dokuchaev (1899) developed the theory of integrated concepts. He pointed out that, within the limits of extensive areas (zones), natural conditions are characterized by many features in common, which change markedly in passing from one zone to another. As S.V. Kalesnik (1962) notes, Dokuchaev "called for the study, not of individual bodies and natural phenomena, but certain integral territorial aggregates of them." These ideas formed the basis for subsequent work in integrated land classification.

At the world scale, "natural regions" have been mapped by Herbertson (1905) (Fig. 1.2), and further refined by Passarge (1929) and Biasutti (1962). In Russia, Berg (1947) coined the term "landscape zones." In Germany, the term *"Landschaft"* is preferred (Neff 1967; Troll 1971). Veatch's (1930) research in Michigan outlined "natural geographic divisions" and "natural land types." In surveys undertaken within the British Empire, Bourne (1931) derived his concepts of "site" and "site regions." Sukachev's investigations into biogeocenology followed similar lines (Sukachev and Dylis 1964). Other studies using integrated concepts have been developed in Australia (Christian and Stewart 1968) and America (Wertz and Arnold 1972) under the title of "land systems." In Canada, such a concept is used in "biophysical" or "ecological

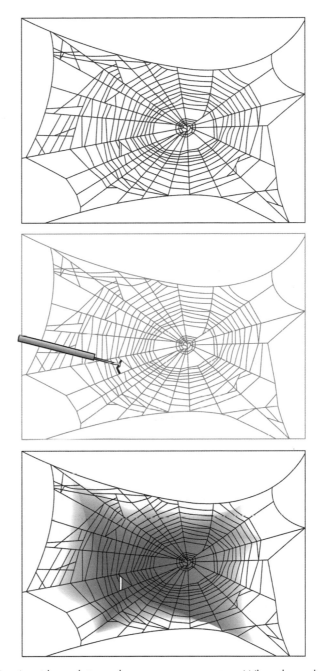

**Figure 1.1.** A spider web is analogous to an ecosystem. When the web is disturbed at one spot, other strands of the web are affected because of linkages.

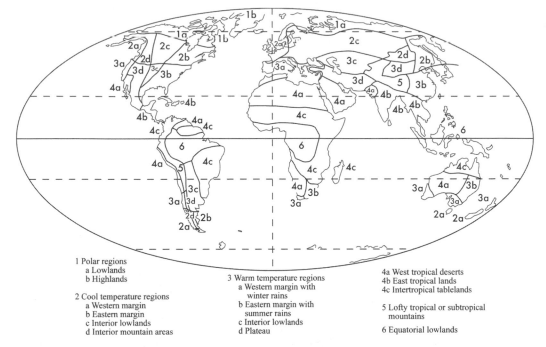

**Figure 1.2.** Major natural regions. From Herbertson (1905).

1 Polar regions
  a Lowlands
  b Highlands

2 Cool temperature regions
  a Western margin
  b Eastern margin
  c Interior lowlands
  d Interior mountain areas

3 Warm temperature regions
  a Western margin with
    winter rains
  b Eastern margin with
    summer rains
  c Interior lowlands
  d Plateau

4a West tropical deserts
4b East tropical lands
4c Intertropical tablelands

5 Lofty tropical or subtropical
  mountains

6 Equatorial lowlands

land classification" (Wiken and Ironside 1977). This methodology calls for total integration of landform, lithology, relief, climate, soils, and vegetation.

Carl Sauer (1925) introduced the term *landscape* into American geography. Geography has progressed in the meantime from the study of landforms, soils, vegetation, and the like to a synoptic consideration of the interrelationships between the elements of nature, independent of their association with a particular place (cf. James 1959; Strahler and Strahler 1976).

# Ecosystem-Based Planning

Optimal management of land ensures that all land uses consistently sustain resource productivity *and* maintain ecosystem processes and function. This equals ecosystem capability; capability provides the context for looking at land-management options. The expression for this relationship is

$$\text{Sustainability} = \text{Resource productivity} + \text{ecosystem maintenance} = \text{Capability}$$

Ecosystem-based planning is the process of prescribing compatible land uses based on capability. The determination of capability requires an understanding of the effects of management practices and prescriptions on the quantity and quality of resource outputs. This, in turn, depends on sound predictions about the behavior of the ecosystem under various kinds and intensities of management, particularly about the effects of management of one resource on another.

# Predicting Effects

The kind and magnitude of expected behavior are the result of many complex and interacting components that control the ecosystem process, such as erosion and vegetative succession. Process is controlled by the ecosystem structure (i.e., how the components are integrated). Various structures and related processes occur throughout any area. Making predictions about ecosystem behavior requires information about the nature of this structure and how it varies geographically.

# Levels of Integration

An ecological map shows an area divided into ecosystems, associations, or integrations of interacting biotic and abiotic features. A method of capturing this integration is the ecological land classification technique (Rowe and Sheard 1981). This technique includes the delineation of unit of land displaying similarities among several ecosystem components, particularly in a way that may affect their response to management and resource production capability. We can show at two levels how these features are associated or integrated. One level shows the integration within the local area, and another shows how the local area is integrated and linked with other areas across the landscape to form larger systems. All these areas are ecosystems, albeit at different scales or relative size. That the ecosystem concept can be applied at any level of spatial scale is suggested by the work of Troll (1971), Isachenko (1973), Walter and Box (1976), Odum (1977), Miller (1978), Mil'kov (1979), Webster (1979), Bailey (1983), Forman and Godron (1986), and Meentemeyer and Box (1987), among others.

# Structure: The Basis of Classification

An inventory of the components of a parcel of land simply provides an inventory of its anatomy; it does not necessarily provide an understanding of how the parts fit together (the structure) and function (Rowe 1961).

How components are integrated at a site, or relatively small area, is called the vertical structure of an ecosystem (Fig. 1.3). However, ecosystems constantly interact with their surrounding systems through an exchange of matter and energy. If we approach ecosystem classification on a structural–functional basis, we must consider both the vertical structure (looking down vertically) of an ecosystem and its interaction with its surroundings. In other words, we must base ecosystem classification on the spatial association of vertical ecosystems. This is the horizontal structure. Setting ecosystem boundaries involves dividing the landscape where the structures exhibit a consistent or significant degree of change when compared with adjacent areas (Fig. 1.4).

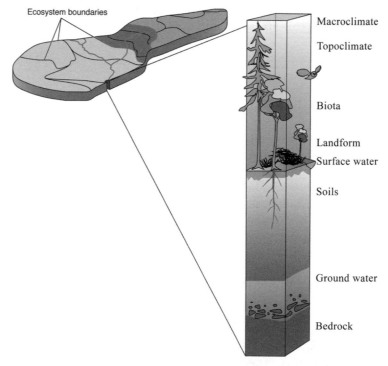

**Figure 1.3.** Vertical structure of an ecosystem.

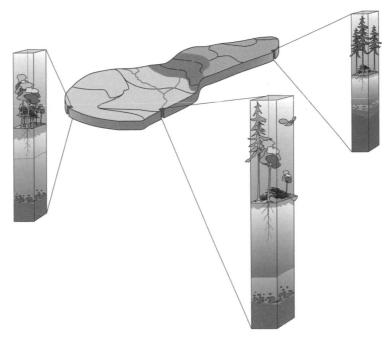

**Figure 1.4.** Boundaries between ecosystems are set where different vertical structures occur.

We can organize information concerning ecosystems by reference to coordinate points. Because by classifying ecosystems we are, in fact, classifying space, point values are of limited value unless we know how they are arranged in relation to their neighbors. We are concerned with conditions that prevail over a given unit area. Ecosystem classification then requires that the characteristics on which the classification is to be based be those of areas. As such, a map is essential to area classification and is indeed the only way to adequately display area location and juxtaposition in a classification system.

In area classification, mapping criteria are defined to establish boundaries where changes in the relationships among area components appear to be most pronounced or significant when compared with adjacent areas. A hierarchy of area classes is formed when areas are grouped together on the basis of association by contiguity. As Rowe (1980) points out, "The key criteria are not to be found simply in the vegetation, in the soil profile, in the topography and geology, in the rainfall and temperature regimes, but rather in the spatial coincidences, patterning and relationships of these functional components." The consideration of *relationships* provides the basis of ecosystem mapping.

# Need for Recognizing Ecosystems at Various Scales

Historically, ecosystems have been defined as small, homogeneous areas or sites, such as a stand of trees or a meadow. There are several reasons for recognizing ecosystems at broader scales as well. Where the boundaries of one ecosystem are entirely enclosed by another's, ecosystems are nested or reside within each other (Fig. 1.5). The boundaries of ecosystems, however, are never closed or impermeable; they are open to transfer of energy and materials to or from other ecosystems. The open nature of ecosystem boundaries is important, for even though we may be dealing with a particular ecosystem as a land unit, we must keep in mind that the exchange of material with its surroundings is an important aspect of the ecosystem's operation.

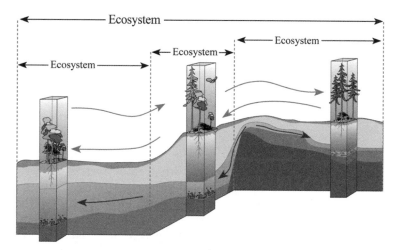

**Figure 1.5.** Ecosystems are nested with permeable boundaries.

Because of the linkages among ecosystems, modification of one system may affect the operation of surrounding ones (Fig. 1.6). Furthermore, how a system will respond to management is partially determined by relationships with surrounding systems linked in terms of runoff, groundwater movement, microclimate influences, and sediment transport. These systems do not exist in isolation. The climate in a meadow is altered by the surrounding forest, for example. We need to work on understanding these linkages so we can better predict the impacts of human activity.

A disturbance to a large ecosystem may affect smaller component systems. For example, logging on upper slopes of one ecological unit may

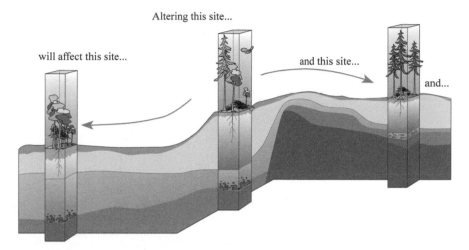

**Figure 1.6.** Effects of alteration of one site on surrounding sites.

affect downslope conditions in smaller nested units, such as stream or riparian habitats (Fig. 1.7). Other forms of vegetation manipulation may have similar effects. For example, chaparral species have deep root sys-

**Figure 1.7.** A meadow surrounded by forest in central Idaho. Sketch by Susan Strawn, from photograph.

**Figure 1.8.** Mouth of Monroe Canyon, southern California: **(a)** before conversion, January 1958; **(b)** photographed from the same position as **(a)** following removal of riparian vegetation; **(c)** after lightning fire of 1960; and **(d)** alluvial accumulation after January and February storms of 1969.

**Figure 1.8.** (*Continued*).

tems and therefore use more water than shallower-rooted grass species. Researchers found that converting vegetation from chaparral to grass to increase the water yield of steep experimental watersheds in southern California affected stream systems through increased discharge rates but also increased debris production (Orme and Bailey 1971). As the roots of the deep-rooted chaparral species decayed, they could no longer anchor the soil on steep slopes. This change decreased the stability of the slopes during storms and increased the amount of material washing downslope. Increased erosion is followed by severe gullying, which in turn is accompanied by aggradation in the main valley. Figure 1.8 depicts this sequence of morphological changes within the drainage basin of Monroe Canyon in the San Gabriel Mountains of southern California. We can extend this concept of system interaction all the way from the smallest watershed to the whole earth.

Because ecosystems are nested spatial systems, each level subsumes the environment of the system at the level below it. Therefore, each ecosystem constrains and controls the behavior of the ecosystem at the level below it (Warren 1979). For example, climate controls runoff in a watershed, which, in turn, interacts with hill slopes to produce stream channels. At each level, new processes emerge that were not present or evident at the next level. As Odum (1977) noted, research results at any level aid the study of the next higher level but never completely explain the phenomena occurring at that level, which itself must be studied to complete the picture. Hierarchy theory (Allen and Starr 1982, O'Neill et al. 1986) is closely related to this idea. A hierarchy is defined as a system of interconnections wherein the higher levels constrain and control the lower levels to various degrees. An important concept from hierarchy theory is the importance of considering at least three hierarchical levels in any study: the level in question, the level above, and the level below.

Some of the processes that are involved in a landscape composed of a mosaic of ecosystems may be in addition to those involved in its separate component ecosystems. They include those processes of interaction among the component ecosystems. For example, a snow-forest landscape includes dark pines that convert solar radiation into sensible heat that moves to the snow cover and melts it faster than would happen in either a wholly snow-covered or wholly forested basin. The pines are the intermediaries that speed up the melting process and affect the timing of the water runoff. Watershed managers can attempt to produce the same effects by strip-cutting extensive forests. Other examples are given by Miller (1978) and Mil'kov (1979).

An example of a smaller ecosystem within a larger controlling ecosystem is a meadow of grass embedded in a forest. It will function differently from a large expanse of grassland. The forest affects the microclimate and the plant cover of the meadow, sheltering the meadow from drying winds or from hail. Many bird species that nest in the forest may feed in the meadow, and meadow rodents like to hibernate at the edge of the forest or in its interior.

At the zones of contact, or *ecotones*, between forest and meadow, the greatest concentration of animal life, mostly insects and birds, occurs. This accounts for the higher density of animal populations in a forest-meadow landscape than in a forest landscape or a grassland landscape (Odum 1971).

In summary, the relationships between an ecosystem at one scale and ecosystems at smaller or larger scales must be examined to predict the effects of management. Because management occurs at various levels, from national to site-specific, one of the prerequisites for rational ecosystem management is to delineate ecosystems at a level, scale, and intensity appropriate to management levels. We therefore need a hierarchical system to permit a choice of the degree of detail that suits the management objectives and proposed use. For a review of the arguments for the recognition of a spatial hierarchy of ecosystems, see Bailey (1985) and Klijn and Udo de Haes (1994).

# Ecosystem Geography

Multiscale analysis of ecosystems pertains to all kinds of land, regardless of jurisdiction or ownership boundaries. Many environmental problems cross agency, state, and national boundaries. These include air pollution, management of anadromous fisheries (fish that go from ocean to freshwater to spawn), introduction of non-native species, forest insect and disease, and biodiversity threats. To address these problems, the planner must consider how geographically related systems are linked to form larger systems. This will require government scientists and managers to integrate their efforts across agency lines. Barriers arise because land-management agencies have disparate missions and user groups. The effect of these different missions is sometimes easily discernible where the lands of these agencies abut one another, as they do along sections of the boundary between Yellowstone National Park, where timber harvesting is prohibited, and the Targhee National Forest in Idaho, where large areas of trees were removed through clearcutting (Fig. 1.9).

A new approach is needed based on *ecosystem geography*, the study of the distribution pattern, structure, and processes of differentiation of ecosystems as interacting spatial units at various scales. As in all branches of geography, emphasis is on the causes behind those patterns. Ecosystem geography is, in many ways, related to the emerging

**Figure 1.9.** Boundary between Yellowstone National Park and Targhee National Forest. Photograph from Greater Yellowstone Coalition, courtesy of Tim Crawford.

field in ecology called "landscape ecology" (cf. Troll 1971; Leser 1976; Forman and Godron 1986). The principal difference between the two is the greater emphasis on mapping in ecosystem geography. A scale difference also exists. Ecosystem geographers have focused greater emphasis on regional and global systems than have landscape ecologists, who, as their name implies, seem to concentrate most of their work at the level below the region (i.e., the landscape level).

Jurisdictional and watershed boundaries will not generally coincide with ecosystem boundaries (Fig. 1.10). We must not restrict ecosystem analysis to the limits of other unassociated boundaries, because we cannot understand an ecosystem by only considering part of it.

Furthermore, we cannot understand ecosystems by only considering their separate components. There is a unity in nature. Ecosystem components cannot function as independent systems, because they exist only in association with one another (e.g., thin soils on steep slopes, flat floodplains of fine-textured soil and inadequate drainage, or the tayga areas dominated by narrow-leaved evergreen forest with Spodosol soil and subarctic climate). We can view how components are related at different levels from the standpoint of the complexity of their relationships. One level provides an understanding of relationships within the local area,

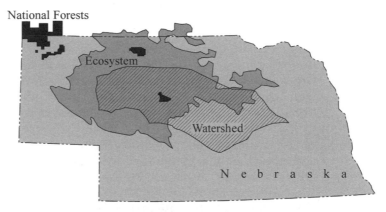

**Figure 1.10.** The Nebraska Sandhills Prairie ecosystem (as mapped by Küchler [1964]) lies partly within the Loup River watershed (as mapped by U.S. Geological Survey [1979]), and vice versa. Jurisdictional forest boundaries and state boundaries have no relationship to the ecosystem.

and another provides an understanding of local areas within the context of a larger area or region.

Integrated classification of small, relatively homogeneous areas is based on their components and involves the combination of two or more components, each with its own hierarchy of levels. For example, we could link a vegetation classification and a soil classification to define ecological units. Combinations could be made from selected levels of the hierarchy in each respective system. The concept of using more than one component of the ecosystem to identify integrated homogeneous units of land at the local level was expressed in the proposed interagency ecological land classification of 1984 (Driscoll et al. 1984). They proposed several component classifications, each with its own hierarchy that can be linked to define ecological land or water units. Integrated units defined in this way are place-independent because interrelationships of surrounding units are not considered. We can group these units on the basis of their similarity into higher classes, which reflects an increasing generality of information. For example, we can group spruce-fir forest ecosystems with Douglas-fir forest ecosystems into a category called needleleaf evergreen forest. Because geographic location is not considered, larger units (higher-level ecosystems) do not necessarily result from such a process. In addition, all data from discontinuous areas of the same type would be pooled regardless of geographic location. This kind of information is necessary to make independent inferences about forest, grassland, and shrubland ecosystems. However, the local ecosystem can never be understood fully except in the context of the larger ecosystem that encompasses it.

For such an understanding, we must view ecosystems in a geographic or spatial hierarchy that reflects how they fit together in the landscape. Grouping ecosystems to define units at this level of integration is analogous to using combinations of soils in defining soil catenas (associations) or landforms in defining watershed basins. However, a problem can arise, because ecosystems related by geography are not necessarily related by taxonomic properties. Taxonomy classifies or groups objects according to similar properties. With geographic units, similarity is not always present. The catena, for example, comprises different taxonomic soil series that are geographically related. Another example occurs where contrasting vegetation types are in juxtaposition because of landform influences on ecosystem patterns (Fig. 1.11).

**Figure 1.11.** Geographically related ecosystems in the semiarid mountains of the Blue Mountains with south-facing grass-covered slopes and north-facing forested slopes. Wallowa National Forest, Oregon. Photograph by Melvin Burke, USDA Forest Service.

An area of spruce forests and glacially scoured lakes constitutes a single landscape ecosystem that is linked internally by downhill flows of water and nutrients, through coarse Spodosol soils, toward clear oligotropic ("few foods") lakes (Fig. 1.12). Geographically related systems such as this, unified by a common mode of exchange of energy and materials, may be combined into larger geographic units referred to as "landscape ecosystems." A landscape ecosystem corresponds closely to the concept of a soil catena, the repetitive mosaic of soil types across a given area.

An advantage of combining ecosystems into larger landscape ecosystems is that we can better relate them to surrounding units with which they interact. This is important in evaluating the effect of management of one type of ecosystem on surrounding ecosystems. For example, we can better evaluate the effect of grazing in the alpine zone on the adjacent subalpine zone. This is in contrast to a taxonomic classification system in which the alpine zones of the Rocky Mountains and Sierra Nevada Range would be grouped because of similar properties, regardless of geographic

**Figure 1.12.** Spruce forests and glacially scoured lakes in Voyageurs National Park, Minnesota. Photograph by Jack Boucher, National Park Service.

proximity. The alpine zones of these two different mountain areas do not interact; they interact with the adjacent subalpine zones in their respective ranges.

# Do We Know Enough?

Some scientists have said that ecosystems are too complex to understand, let alone to manage. Yet, ecosystems have been managed for centuries with imperfect knowledge. Today, we have amassed great volumes of information about ecosystems: so much so that information overload has become a problem. We need a synthesis of available information and the ability to apply it to management. Work is needed not in presenting information by itself but in striving for synthesis (i.e., the illustration of interrelationships).

We approach "truth" by a series of approximations. For example, the U.S. Geological Survey has been producing geologic maps of the nation for more than a century. Every few years during this period a new map has been published, each somewhat different from the previous edition. Does this mean that the geology has changed? No, it means that the geologist's understanding of the geology has changed and improved, creating the need for a new map. The same concept applies to ecosystem classification, mapping, and management. We must use the best tool for management that our current understanding will permit, recognizing that the products of these efforts will be updated and improved in the future as we learn more.

# Need to Delineate Ecosystem Boundaries

Ecosystems and their components are naturally integrated. They existed before mankind appeared and would continue to exist if mankind disappeared. In other words, we do not integrate anything; it is already integrated. The task of the ecological land mapper is to understand and capture that integration. Unfortunately, there is disagreement on how many ecosystems to delineate and what specific criteria to use to separate one system from another.

Another problem with setting boundaries is that most natural components of an ecosystem, which might be used in defining it, vary along a continuum. The boundaries, therefore, must often be defined as a zone of transition and may be arbitrary or indistinct. This does not diminish their value, however. Generalization is an integral and inescapable part of all mapping. As a result, mapped units of most kind will vary in purity and uniformity. For example, the Pierre Shale shown on a geologic map does not consist of shale throughout, laterally and in vertical section.

Generalization processes may involve simplifying boundaries or allowing atypical conditions to be included in the map unit. The degree to which this occurs will partly depend on the scale of the map as well as its purpose. The boundary of a small-scale map may be considerably different in detail than that of the large map in which the area resides.

Map scale aside, maps may show different sizes of ecosystems. This can explain some of the differences among ecosystem maps by different authors. The patterns of ecosystem boundaries on these maps may be different because they are aimed at differentiating ecosystems of different rank. One map may depict large ecosystems; another, the smaller ecosystems that may exist within the larger. For example, we can map a pattern of combined component systems or map the individual component systems themselves. At first observation, these two maps may appear contradictory. They are not. They are simply different but compatible expressions of the same phenomena (Fig. 1.13).

These facts do not negate the need to delineate ecosystem boundaries. They are prerequisite to mapping for purposes of analyzing and managing ecological units and land use. We can delineate boundaries so they define ecosystems for general purposes and as a starting point for more specific purposes. To accomplish this, classification should be based on the following principles:

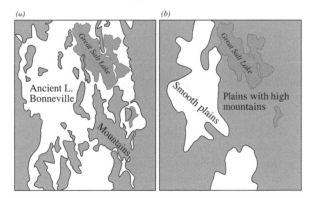

**Figure 1.13.** Maps may express different interpretations of the same phenomena without being contradictory: (**a**) a portion of the Basin and Range area in Utah bordering the plain of ancient Lake Bonneville; (**b**) the same area in which the mountain ranges are not differentiated from the plains.

1. The system should be based on multiple factors. Ecosystems are defined by multiple factors (both abiotic and biotic). As Sokal (1974) points out, "Classifications based on many properties will be general: they are unlikely to be optimal for any single purpose, but might be useful for a great variety of purposes." This is termed a *natural classification*.

2. The system should be based on causes. A fundamental principle of scientific classification is that establishing classes of things is better done according to the *causes* of the class differences than according to the *effects* that such differences produce (Strahler 1965). The units derived from such a classification are termed *genetic*. As Rowe (1979) points out, the key to the placing of map boundaries on ecological maps is the understanding of genetic processes. We can only comprehend a landscape ecosystem if we know how it originated. That is why Huggett (1995) suggested that the approach is evolutionary as well.

# The Genetic Approach

The genetic approach looks for patterns in the landscape and seeks to understand the formative processes that create those patterns. For example, trees that respond to additional moisture on north-facing slopes or along streams are seen repeatedly throughout semi-arid and arid regions of the American West. These patterns are not isolated occurrences but are inextricably linked to the ecological processes that shape them. Repeated patterns emerge at varying scales. For example, temperate steppes in the Northern Hemisphere are always located in the interior of continents and on the windward, or western, sides; thus, the central United States is in some ways similar to steppes in Eurasia, the pampas in South America, and the veldt in Africa. In the Denver, Colorado area, the rocky forested Front Range slopes with a typical sequence or spectra of altitudinal belts which rise abruptly from the grassy plains are among the most prevalent patterns in that region. Rocky outcrops on the nearby Great Plains grasslands are repeatedly accompanied by islands of trees and shrubs that tap into associated reservoirs of water. Thus, the genetic approach is the act of understanding the patterns of a region or a site in terms of the processes that shape them and then applying these to differentiate the landscape into ecosystems of various scales. For practical application, understanding spatial relationships between causal mechanisms and resultant patterns is a key to understanding how ecosystems respond to management.

The notion that process knowledge should define landscapes has a long history in geography. William Morris Davis (1899) argued for a genetic classification of landforms, and both Herbertson's (1905) natural regions of the world (Fig. 1.2) and Fenneman's (1928) physiographic divisions of the United States (Chapter 3) are based on rational observation of underlying processes. Likewise, both Dryer (1919) and Sauer (1925) called for a genetic approach to landscape classification.

The following chapters present an approach for applying the principles of multiple factors and genetic process that are useful for delineating and managing ecosystems for multiple purposes at several geographic scales.

# CHAPTER 2

# Scale of Ecosystem Units

$S$cale implies a certain level of perceived detail. Suppose, for example, that we carefully examine an area of intermixed grassland and pine forest. At one scale, the grassland and the stand of pine each appear spatially homogeneous and look uniform. Yet linkages of energy and material exist between these ecosystems. Having determined these linkages, we intellectually combine the locationally separate systems into a new entity of higher order and greater size. These larger systems represent patterns or associations of linked smaller ecosystems.

Several countries have proposed and implemented schemes for recognizing such scale levels (Table 2.1; see also Zonneveld 1972; Salwasser 1990; Klijn and Udo de Haes 1994; Blasi et al. 2000). In these schemes, the nomenclature and number of levels vary. One scheme, proposed by Miller (1978), recognizes linkages at three scales of perception. Rowe and Sheard (1981), although using different terminology, advanced a similar scheme (Table 2.2). A few years later (Bailey 1985, 1987, 1988a), I proposed a hierarchical ecosystem classification inspired by both of these schemes and closely following Miller's terminology. It is the framework for this book. A hierarchy of ecosystem units based on this framework is illustrated in Figure 2.1.

## Site

The smallest (a few hectares), or local, ecosystems are the homogeneous *sites* commonly recognized by foresters and range scientists. We refer to these as microecosystems.

R.G. Bailey, *Ecosystem Geography*, DOI 10.1007/978-0-387-89516-1_2,
© Springer Science+Business Media, LLC 2009

**Table 2.1.** Comparison of the nomenclature of some ecological classification systems of hierarchical character—comparable concepts have been placed on the same level[a]

| Australia | Britain | Canada | USSR | United States |
|---|---|---|---|---|
| | | | Zone | |
| | | | | Domain |
| | Land zone | | | Division |
| | Land region | Ecoregion | Province | Province |
| | Land district | Ecodistrict | | Section |
| | | | Landscape | |
| Land system | Land system | Ecosection | | District |
| | Land type | Ecosite | Urochishcha | Landtype association |
| Land unit | | | | |
| Land type | Land phase | | | Landtype |
| Site | | Ecoelement | | Landtype phase |
| | | | Facia | Site |

[a]From Bailey (1981).

**Table 2.2.** Levels of generalization in a spatial hierarchy of ecosystems

| | Scheme | | |
|---|---|---|---|
| Miller (1978) | Rowe and Sheard (1981) | Approximate size (km$^2$) | Map scale for analysis |
| Region | Macroecosystem | $10^5$ | 1:3,000,000 |
| Landscape mosaic | Mesoecosystem | $10^3$ | 1:250,000–1:1,000,000 |
| Ecosystem (site) | Microecosystem | 10 | 1:10,000–1:80,000 |

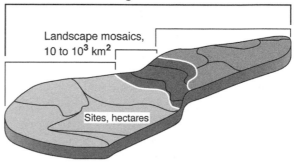

Ecoregion, $10^5$ km$^2$

Landscape mosaics, 10 to $10^3$ km$^2$

Sites, hectares

**Figure 2.1.** Hierarchy of ecosystems.

# Landscape Mosaic

Linked sites create a *landscape mosaic* (mesoecosystem), or simply landscape, that seen from above looks like patchwork. A landscape mosaic is made up of spatially contiguous sites distinguished by material and energy exchange between them. They range in size from 10 km² to several thousand square kilometers.

A mountain landscape is a classic example of a landscape mosaic. A lively exchange of materials occurs among the component ecosystems of a mountain range: water and products of erosion move down the mountains; updrafts carry them upward; animals can move from one ecosystem into the next; seeds are easily scattered by the wind or distributed by birds.

# Ecoregion

On broader scales, landscapes are connected to form larger units (macroecosystems). Mountains and plains illustrate this well (Fig. 2.2). For

**Figure 2.2.** Ecosystems can be considered at various scales. In this view of Death Valley in California, the *macroscale* is represented by the mosaic of deeply eroded ranges and smooth basin floors. The *mesoscale* is represented by the two components of the mosaic—ranges and basins. The *microscale* is represented by individual slopes within the mountain ranges. Photograph by Warren Hamilton, U.S. Geological Survey.

example, as a mosaic, the lowland plains of the western United States contrast with steep landscapes in adjacent mountain ranges. As water from the mountains flows to the valley and as the mountains affect the climate of the valley through sheltering, two large-scale linkages are evident. Such linkages create real economic and ecological units. This unit is called an *ecoregion*, or simply region. Regions occur in many scales (Bailey 1983). Like landscapes, they stand in contrast with one another, while long-distance linkages connect them. Finally, this progression reaches the scale of the planet.

# National Hierarchy of Ecological Units

Recently, the U.S. Forest Service (ECOMAP 1993) more elaborately followed the ideas presented above. Instead of three levels to be distinguished, they recognized more levels but on the same principles (Fig. 2.3). In 1993, the agency adopted this hierarchy for use in ecosystem management.

I mapped ecoregions down to the province level for the United States (Bailey 1976, revised 1994), North America (Bailey and Cushwa 1981, revised 1997), and the world's continents (Bailey 1989). Bailey et al. (1994) mapped the ecoregion subregions or sections of the United States. Cleland et al. (2005) developed a map of the conterminous United States showing subsection boundaries as well as another approximation of section boundaries. This map was compiled from subsection maps of each Forest Service region (cf. Nesser et al. 1997). Many national forests in different parts of the country have produced maps of landtype associations.

**USDA Forest Service**

| Bailey 1988a | Ecomap 1993 | |
|---|---|---|
| Ecoregion | Ecoregion | Domain |
| | | Division |
| | | Province |
| Landscape Mosaic | Subregion | Section |
| | | Subsection |
| | Landscape | Landtype Association |
| Site | Land Unit | Landtype |
| | | Landtype Phase |

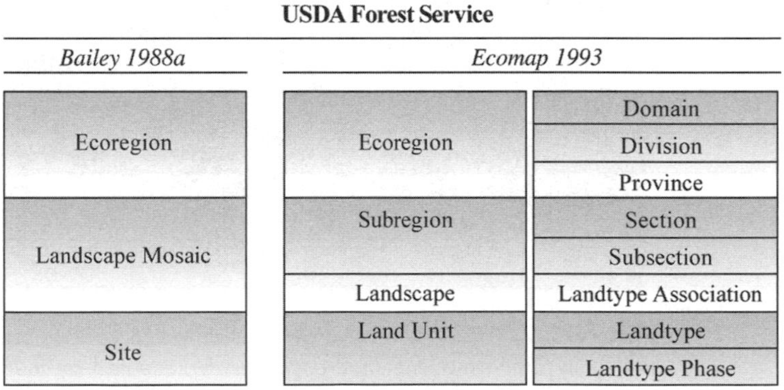

**Figure 2.3.** Comparison of hierarchies used for ecological land classification in the U.S. Forest Service.

# CHAPTER 3

# The Question of Boundary Criteria

Two fundamental questions facing all ecosystem mappers are the following: (1) What factors are of particular importance in the recognition of ecosystems? (2) How are the boundaries of the different sizes of systems to be determined?

We can use five basic methods to identify land units where ecosystem components are integrated in similar way, thereby classifying land as ecosystems: gestalt, map overlay, multivariate clustering, digital-image processing, and controlling factors.

## Gestalt Method

A gestalt is a whole that is not derived through consideration of its parts. The gestalt method recognizes homogeneous-appearing regions and draws boundaries intuitively, based largely on visual appearance in the field or on aerial photographs or satellite imagery. This method generally does not consider individual factors such as slope, soils, and vegetation (Hopkins 1977). An area is partitioned by implicit judgment—rather than on explicit rules—into so-called homogeneous regions, such as uplands or lowlands. The philosophy of this method seems to be that no rules exist for recognizing regions; they vary depending on location. Geologic structure and relief, for instance, are guiding factors in the delineation of major systems such as the Rocky Mountains, whereas low rainfall delineates the Great Plains. These schemes eventually evolve into nothing more than "place-name regions," which are identified primarily by the places themselves rather than by objective criteria that define particular types of regions.

R.G. Bailey, *Ecosystem Geography*, DOI 10.1007/978-0-387-89516-1_3,
© Springer Science+Business Media, LLC 2009

Regions generated without identifying which factors were considered are difficult for others to scrutinize or confirm. The results are therefore difficult to communicate convincingly to decision makers. We need more explicit methods. Such explicit methods inherently require considering factors that enter into ecosystem differentiation. The remainder of this chapter is concerned with how we can explicitly combine such factors to yield ecosystem maps.

# Map-Overlay Method

On the premise that many maps of factors pertinent to ecosystem definition may be interrelated and therefore correlated to each other, a method of overlaying maps is thought to have potential, by some, for identifying zones where factor boundaries correspond to each other. The perceived areas are thought to define ecosystems, also called ecological response units. Typically, this method relies on available maps only.

Although map overlaying may be useful in identifying ecosystems, the approach has shortcomings (Bailey 1988b; Lowell 1990). To begin with, boundaries on different factor maps rarely correspond to each other (Fig. 3.1). This is because each factor has been studied independently by different professionals at different times with different purposes in mind.

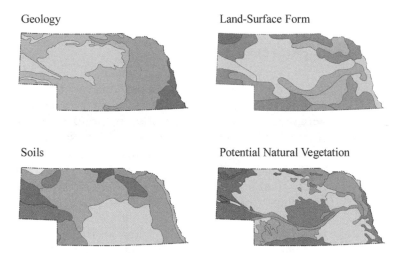

**Figure 3.1.** Maps of natural factors that might be considered in defining ecosystem units in Nebraska. Geology from Kinney; land-surface form from Hammond; soils from Soil Conservation Service; potential natural vegetation from Küchler. From U.S. Geological Survey (1970).

Different principles and methods, degrees of detail, and errors in source maps combine to detract from an integrated ecological picture.

Boundaries of ecological significance will, instead, emerge from studies that reveal corresponding changes in the natural factors. An example would be studies that focus on the zoning of vegetation in response to change in geology. This is different from attempting to synthesize ecosystem units by addition of factors or components initially defined as things in themselves, with no whole unit in mind. Moreover, the problems of boundary-line location and impurity of mapping units generated by overlaying maps create some time-consuming, if not impossible, difficulties. These arise from the need to interpret and overcome the erroneous factor combinations that result from trying to combine independently derived information.

Additionally, using available factor maps may not work well for identifying ecosystem units. First, the same factor may indicate different process rates, depending on where it is observed. For example, studies have shown that the productivity of the same soil series varies considerably throughout its range (Gersmehl 1980). Second, factor maps reflect a classification. The class boundaries selected for the map may not be relevant to the initiation of a land process. For example, the slope angle that indicates an erosion threshold for a particular geologic material varies, depending on the regional climate where the slope is located. Slope maps typically do not account for this variation.

Another problem with the map-overlay method is lack of information. Some factors that may be critical to understanding a process do not exist in map form. For example, the degree to which a land surface is dissected by streams is critical to understanding the process of sediment transport. But this information does not commonly exist in map form showing various dissection classes. Although analysis could obtain such information, it is rare because using available maps is a practical necessity.

The rationale behind the map-overlay method is the notion that significant ecosystem units can be captured by synthesizing, or integrating, available factor maps. The implied assumption is that the derived units reflect differences in potential response to management and resource productivity. It has often been assumed without validation that the commonly synthesized factors are the most appropriate for expressing these differences. As yet, this synthesis has developed little beyond an empirical description that provides no explanation of processes that produced the units identified. This limits the ability to predict productivity and the consequences of environmental impact. To be effective, such an environmental synthesis must be shown to apply directly to process (see the review by Moss 1985).

# Multivariate Clustering Method

Similar problems are encountered in the application of multivariate clustering to classify land by grid cells (Omi et al. 1979). In this approach, an arbitrary grid of cells is imposed on the surface to be mapped. The cells are then described by selected attributes and the information entered into a geographic information system. This information is then used to classify cells by numerical taxonomic methods, such as cluster analysis. A map is produced by drawing lines around cells of similar class. However, as Rowe (1980) points out, the units derived from such a process are not necessarily ecological. Ecological units can be comprehended only as wholes that have some process significance. For example, a floodplain is a pattern of spatially associated, but *unlike* component land units (cells). The floodplain consists of the active channel, abandoned channels, islands, lakes, wetlands, levees, and so forth (Fig. 3.2). Each unit has different characteristics but is united with the others by common processes of development, namely, cyclic inundation, erosion, meandering, and deposition.

**Figure 3.2.** The braided channel of the Rio Grande in northern New Mexico. During flood, the entire belt of channels and sandy islands will be covered with water. Photograph by E.D. Eaton, Soil Conservation Service.

# Digital-Image Processing Method

A related approach, with the same limitations, is digital-image process-ing (Robinove 1979). In this sophisticated grid approach, the cells are very small. Maps are created directly from satellite or photo imagery. Again, clustering of cells on the basis of their appearance from the imagery does not necessarily result in the identification of ecological units. This is because cells with different spectral signatures frequently occur in the same ecosystem.

# Controlling Factors Method

Some scientists faced with the staggering complexity of boundary pat-terns when using the map-overlay approach have concluded that some form of simplification is necessary. In other words, the number of fac-tors must be reduced. As an alternative to the overlay approach, the con-trolling factors method is based on the dominance or greater relevance of one particular environmental controlling factor. With this approach, certain key factors are recognized to exert a strong influence on the ecological process of the land, and hence on resource management. These factors are used to partition the landscape into ecological units for planning analysis at different spatial scales. The following section reviews the logic and criteria for subdividing a landscape based on this approach.

The ecological units that are derived by the controlling factors method could be used as a layer with other factor maps. This layer, defined in terms of process, would constitute the basic ecological framework for analysis which we can then describe by reference to the other layers. For example, we could use hierarchical classifications for soil and vegetation to describe ecosystem unit composition (Table 3.1).

Because we can understand subsystems only within the context of the whole, a classification of ecosystems usually begins with the largest units and successively subdivides them by levels. Although the concept of ecosystem implies equality among all components, all components may not be equally significant throughout the hierarchy. Further, we cannot possibly consider all these components at the same time. When subdi-viding them into even smaller units, we must prioritize each component to reflect its level of control on the location, size, productivity, structure, and function of the system. Thus, components that exert the most control are highest in the classification.

The differentiating criteria at the upper levels are broad and general in importance with the greatest control, whereas those at lower levels

**Table 3.1.** Examples of hierarchical classification systems used to describe ecosystem unit composition

| Ecosystem unit scale | Soils[a] | Potential vegetation[b] |
|---|---|---|
| Macro | Order (Mollisol) | Class (forest) |
| | Suborder (Boroll) | Subclass (coniferous forest) |
| | | Formation (Temperate Mesophytic forest) |
| Meso | Great group (Cryoboroll) | Series (grand fir) |
| Micro | Family (clayey, Lithic Cryoboroll) | Plant association (grand fir/ginger) |
| | Phase of family (eroded phase) | Ecological site (sandy substrate phase) |

[a]Taxa presented follow Soil Taxonomy (USDA Soil Conservation Service 1975).
[b]Taxa presented follow Driscoll et al. (1984).

are narrow and more specific in importance. Climate zones, for example, determine the global patterns of ecosystems. Soils show more fine-grained patterns, whereas vegetation superimposes an even finer pattern of various succession stages. We may observe that the various components of the ecosystem are ecologically relevant at different time scales, as can be understood by looking at the natural rate of change in the components. Climate changes generally take tens of thousands of years. Vegetation, however, may react within 1 year, whereas fauna is the most rapidly responding component of the ecosystem. These differences in the temporal scale of natural processes reflect the level at which they are most important in the hierarchy. The most rapidly responding components are put relatively low in the hierarchy. Thus, integrated classification of ecosystems must be concerned with a range of components based on a good understanding of the controlling processes involved for the differentiation of successive levels.

Many possibilities for primary factors are apparent, such as vegetation, soil, and physiography. Within each possibility, we can make several other subsequent choices. We look at these choices in detail below.

## Vegetation and/or Fauna as a Basis for Ecosystem Delineation

Vegetation maps may record the current nature of the vegetation, indicating such features as dominant species and height of canopy, as well as vegetation at various stages of succession, and areas where the vegetation has been cleared. Alternatively, maps may present the potential vegetation of the area (i.e., the climax vegetation likely to be present in the

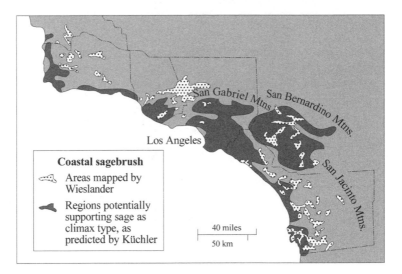

**Figure 3.3.** Differences between maps of existing and potential vegetation in southern California. From Westman (1985), p. 206. From Westman. *Ecology, Impact Assessment, and Environmental Planning.* © 1985 by John Wiley & Sons, Inc.; reprinted with permission of John Wiley & Sons, Inc.

absence of human interference, given the climate, soil, and topography of the region) (Fig. 3.3).

Vegetation and associated fauna, or biota, are constantly changing due to disturbance and succession. For example, fires or timber harvesting may destroy a forest, causing fauna dependent on the forest to migrate. As the process of succession restores the forest to predisturbance conditions, the fauna will repopulate the forest. Moreover, the biogeographic distribution of animal species or communities may change due to hunting, independent of habitat loss. The distribution of bison (*Bison bison*) in North America is a good example of this (Fig. 3.4).

We need to base ecosystem boundaries on the factors that control ecosystem distribution at various scales rather than on present biota to screen out the effects of disturbance or natural successions. This way, ecosystems can be recognized, compared, and worked with regardless of the present land use or other disturbance. The potential of any system makes it possible to understand or manage it wisely.

## Soil as a Basis for Ecosystem Delineation

Even less satisfactory is the use of soil types as the basis for a major subdivision. Soil profiles can rarely be seen except in road cuts. They are usually determined by sampling through drilling or excavating. This

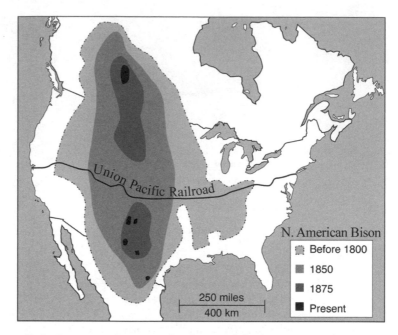

**Figure 3.4.** Former and present distribution of the bison in North America. From Ziswiller in Illies (1974), p. 95; reproduced with permission from Macmillan Press Ltd.

puts a practical limit on the number of samples that can be acquired in an area that is to be mapped. Soil maps, therefore, are usually made by correlating soil samples to other components such as landform and vegetation, which act as surrogates for soils. Soil type frequently does not reflect climate, because the nature of the geologic substratum influences the profile. In tropical areas, soil profiles are usually extremely old and show characteristics that were established under quite different climatic conditions from those now prevailing. There also are many "fossil" soils. The "terrarossa," for example, was once regarded as the typical profile for an area with a Mediterranean climate, but today it is considered to be a fossil tropical soil (Walter and Breckle 1985). Many other soils are relics of some other climatic regime.

## Physiography as a Basis for Ecosystem Delineation

In its original sense, the term *physiography* is a contraction of physical geography, which is the study of the features and nature of the earth's surface, atmosphere and climate, distribution of plant and animal life, and so on (cf. Bowman 1911; Joerg 1914; Atwood 1940). However, a geologist produced one of the best-known physiographic maps of the United States. The task of preparing the map was entrusted by the Physiographic Committee of the Geological Survey to N.M. Fenneman, and his physiographic map was published in 1914 (Fenneman 1928). This map was based largely on structural geology (e.g., the Ridge and Valley province), although certain landform attributes, notably relief and degree of dissection, were also used. Where major physiographic discontinuities occur, where mountains meet plains, or where igneous rocks change to sedimentary strata, the boundaries of these units often coincide with changes in the biota. In areas of little relief, such as the Great Plains, there tends to be little or no correlation of the geologist's concept of physiography with ecology.

As we see in Chapter 4, solar energy plays a major role in ecosystem differentiation. Latitudinal position has a greater effect on the controlling climate than does geologically based physiography. As a result, many times the boundaries of such physiographic units cut across energy zones and their associated ecosystems. For example, the northern Great Plains in Canada will have a considerably different climate than the southern Great Plains in Texas. Therefore, the magnitude of the influences that physiography/substrate has on ecosystems also varies with latitude. Physiography appears to modify the climate within a latitudinal zone and therefore has a secondary effect on ecosystem differentiation.

## Watersheds as a Basis for Ecosystem Delineation

Some problems with using watersheds for defining ecosystem boundaries may be identified. First, large areas (perhaps as much as 20% of the United States) do not have clearly defined drainage networks; these include deserts, the wetlands of Florida and the Lake States, and northern glaciated areas of prairie potholes. Determining just where the water is flowing in such areas is problematic.

Second, a watershed is usually defined by surface-water drainage bounded by a "surface-water divide" that coincides with topography. The associated groundwater divide does not necessarily coincide with the surface-water divides nor will groundwater movement necessarily be parallel to the flows of the rivers and streams. The interrelationships of the surface water and the groundwater are an integral part of the dynamics of hydrologic systems. Therefore, even where the surface-water flow is definable, the associated hydrologic system would be difficult to delineate on a map.

Third, a watershed's stream system may flow through areas of quite diverse climate and landform. For example, the South Platte River in Colorado begins in the high basins of the Rocky Mountains and flows through the rugged Front Range and into the Great Plains. This river reacts to different environments with accordingly different characteristics. The streams flowing into this river have very different thermal characteristics (Fig. 3.5), gradients, aeration, and resultant biota. Conversely, Cateau du Missouri, in the Temperate Prairie Parkland, a climatic-landform unit, drains partly into the Missouri River Basin and partly into the Red River of the north. We would expect that streams throughout those units would have a degree of similarity, regardless of which river they drain into.

Similar climatic-landform units define similar kinds of aquatic environments, but watersheds, particularly the large ones, do not capture those similarities. Note that the biota of aquatic environments may not be similar if the drainage networks are not integrated (connected), blocking migration or if disturbances such as acid-mine drainage have changed the biota.

As Omernik and Griffith (1991) point out, "While river basin units are appropriate for some types of hydrologic data, rarely do the spatial differences in the quality and quantity of environmental resources correspond to topographic divides."

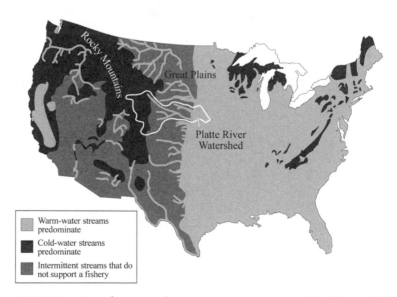

**Figure 3.5.** Location of streams that support warm-water and cold-water fish and streams that do not support fishery. From Funk (1970), p. 142; reproduced with permission from American Fisheries Society.

## Aquatic Biota as a Basis for Ecosystem Delineation

Aquatic biota are dependent on the watershed *characteristics*, which are determined by the ecosystem in which they reside. As they change, so do the biota. Where a watershed flows through more than one ecosystem, the biota distribution may be more related to the ecosystem than the boundary of the watershed itself. One example, above, of the South Platte River applies here. The associated fish population of the Rocky Mountains greatly differs from that of the Great Plains. Trout thrive in the high mountains, whereas other species are predominant in the Great Plains.

Human influences may have altered biota from a natural state. Communities of biota are dynamic and respond to many complex factors and are subject to rapid change; for example, the vegetation associated with the Prairie Pothole region is temporarily very unstable (Lew Cowardin, written communication, 1993).

In addition, various communities of biota may overlap but not coincide.

Considering all the above, we can see that ecosystem maps based on the biogeographic distribution of aquatic or terrestrial species alone do not define an ecosystem.

# Analysis of Controlling Factors

Some argue that not one but many ecosystems may exist in any given place, depending on the viewpoint of the analyst. Such a philosophy creates ecosystems with ever-changing shapes like an "amoeba." Different criteria produce radically different maps for the same area (Fig. 3.6). Others argue that ecosystems are constantly in a state of flux and therefore cannot be delineated by fixed boundaries. For example, ecosystem patterns result, in part, from variability in climate and landforms. Disturbance and the subsequent development of vegetation, as well as human land use impose patterns on these patterns that change through time. In addition, interactions among organisms, such are competition and predation, may lead to other spatial patterns, even in the absence of abiotic variations. The lack of a common fixed unit would make multidisciplinary research and management very difficult, if not impossible. The lack of a common unit for analysis also makes dealing with resource interaction difficult.

Again, one approach to solve this problem is to analyze those factors that *control* ecosystem size at varying scales in a hierarchy and to use the significant changes in those controls as the boundary criteria. This

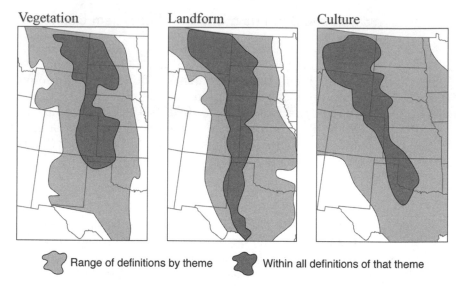

Figure 3.6. Alternative definitions of the Great Plains. Each map is based on the definitions given by seven leading authorities. From Lewis (1966), pp. 142–143. From Lewis. Regional ideas and reality in the US-Rocky Mountain West. *Trans. Inst. British Geog.* 38; reprinted with permission of Blackwell Publishing.

will screen out the effects of disturbance or plant succession, permitting identification regardless of what currently exists. We can then identify permanent mappable ecosystem boundaries.

As discussed in Chapter 1, large ecosystems are made up of dissimilar component systems. To show links between systems and to establish a hierarchy, they should be based on attributes common to all levels. *Climate* is the common attribute and prime controlling factor. We discuss the role of climate in ecosystem differentiation in the next chapter.

# CHAPTER 4

# Role of Climate in Ecosystem Differentiation

Climate is the composite, long-term, or generally prevailing weather of a region. As a source of energy and water, it acts as the primary control for ecosystem distribution (Fig. 4.1).

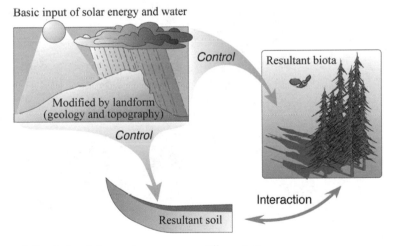

**Figure 4.1.** Role of climate in ecosystem differentiation.

Landform plays a critical role in the modification of climate, and visa versa. This combination of landform and climate controls the patterns in dependent components, such as soil and biota. In other words, the soil and biota are a function of climate and landform. The expression for this relationship is

$$\text{Soil and biota} = f(\text{Climate and landform})$$

R.G. Bailey, *Ecosystem Geography*, DOI 10.1007/978-0-387-89516-1_4,
© Springer Science+Business Media, LLC 2009

As the climate changes, the other components of the system change in response (Fig. 4.2). As a result, ecosystems of different climates differ significantly. Climate would appear to be, then, the initial criterion in defining ecosystem boundaries. Modifying landform should be next, with other criteria following. We discuss the validity of this premise in the following chapters.

These climatic differences result from factors that control *climatic regime*, defined as the diurnal and seasonal fluxes of energy and moisture. We can illustrate different climatic regimes by studying climate diagrams, or climographs. For example, tropical rainforest climates lack seasonal periodicity, whereas midlatitude steppe climates have pronounced seasons (Fig. 4.3).

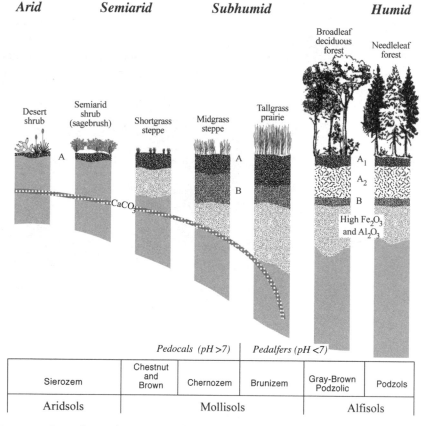

| | Pedocals (pH >7) | | Pedalfers (pH <7) | | |
|---|---|---|---|---|---|
| Sierozem | Chestnut and Brown | Chernozem | Brunizem | Gray-Brown Podzolic | Podzols |
| Aridsols | Mollisols | | | Alfisols | |

**Figure 4.2.** Relationships among climate, vegetation, and soils along a line from the arid Southwest to the Great Lakes region of the United States. From Bear et al. (1986) in Akin (1991), p. 256. From Akin. *Global patterns: climate, vegetation, and soils.* © 1986 by the University of Oklahoma Press; reprinted with permission from University of Oklahoma Press.

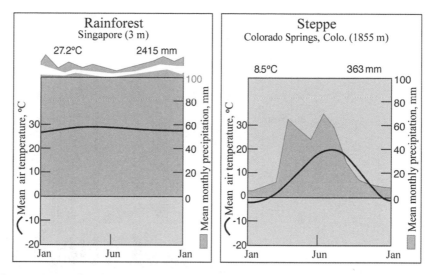

**Figure 4.3.** Climatic regimes as a basis for ecosystem differentiation. Redrawn from Walter et al. (1975).

We adapted the climate diagrams used in this book from the well-known system of Heinrich Walter (Walter and Lieth 1960–1967; Walter et al. 1975). In the climate diagrams, one division on the vertical axis is equivalent to 10°C or 20 mm precipitation. The curves give the mean monthly values of temperature and precipitation, and the scale ratio of 10°C = 20 mm rain (i.e., 1:2 holds for all diagrams). The temperature curve in relation to the precipitation curve is used instead of a potential evaporation curve, for which measured values are available only from a few stations. The occurrence of a relatively dry season is depicted by placing the precipitation curve below the temperature curve (Gaussen 1954).

On the top of each diagram is the location of the weather station, its elevation (in parentheses, in meters), the average annual temperature (°C), and the average annual precipitation (mm).

# Hydrologic Cycle

As the climatic regime changes, so does the hydrologic cycle (Beckinsale 1971). Streamflow is one component of the hydrologic cycle we can measure to help us define a climatic regime. The hydrographs of daily streamflow for three small rivers are shown in Figure 4.4. Streamflow is highest in winter and spring for all three, when evaporation rates are low and soil moisture and groundwater supplies are greatest, and lowest in late summer and autumn, when evapotranspiration

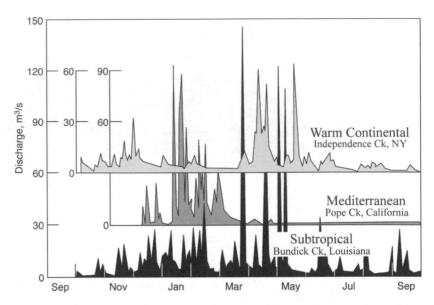

**Figure 4.4.** Hydrographs for three small rivers in different climate regions. Adapted from Muller and Oberlander (1978), p. 166; reproduced with permission.

rates are high. Peaks in the hydrographs are due to surface runoff from precipitation.

The rivers are located in different climatic regions. Bundick Creek in Louisiana is representative of a warm, humid subtropical climate.

**Figure 4.5.** Landforms of two different climatic regions: semiarid, Colorado Plateau, Arizona (*left*); humid, Appalachian Plateau, Pennsylvania (*right*). Photographs by George A. Grant, National Park Service (*left*) and USDA Forest Service (*right*).

Streamflow is greatest during winter and spring; the peaks in summer and fall are associated with runoff from heavy thunderstorms. Independence Creek is in the humid but cold climate of the Adirondack Mountains region of New York. It has a warm-summer, continental climate. Runoff normally peaks in early spring due to snowmelt and spring rains; a secondary peak occurs in late autumn because of decreasing evapotranspiration. The minimum flow in winter is associated with winter snow cover and brief winter thaws. By contrast, no water flows in Pope Creek, located in a warm, dry, summer region of California, during summer and fall, but in winter and early spring, groundwater contributes to streamflow. Additional discussion of hydrology and other relevant ecosystem processes are presented in Chapter 11.

## Landforms and Erosion Cycles

Climate also profoundly affects landforms and erosion cycles (cf. Tricart and Cailleux 1972). Figure 4.5 shows how slopes formed by erosion will vary when similar rock types are exposed to different energy and moisture conditions. These views illustrate variations in slope form in physically similar, horizontally layered sandstones and shales. In the

**Figure 4.5.** (Continued).

Colorado Plateau in Arizona (left photo), the steeper cliffs are formed by the more resistant sandstone formations, whereas the gentler slopes (approximately 30°) are formed by the softer, less resistant shale layers. This arid climate, with only about 150 mm of precipitation, supports little or no vegetation or soil to protect the bedrock. Rapid runoff and flash flooding from thunderstorms carry scouring material that cuts into the land surface, forming gullies and shedding coarse waste onto the slopes.

Similar bedrock, however, in the Appalachian Plateaus in Pennsylvania (right photo), is protected by forest cover and deep soil developed in response to a wet climate with about 900 mm of annual precipitation. Most of the precipitation is snow and spring rains that do not produce flooding. The deep soils increase water infiltration into the ground, which also reduces erosive surface runoff.

# Life Cycles

Plants and animals have adjusted their life patterns to the basic environmental cycles produced by the climate. Whenever a marked annual variation occurs in temperature and precipitation, a corresponding annual variation occurs in the life cycle of flora and fauna. During the cold or dry season, most insects become dormant or die, leaving eggs or larvae behind to hatch in the following, more favorable season. Some higher mammals and birds hibernate; others migrate.

Annual cycles are very apparent in the tropical grasslands and in midlatitudes. The rainforest and polar deserts are about the only regions that do not experience annual changes. Many plants and animals adjust to the changing length of day throughout the year. Reproduction, dormant periods, color changes, migration, and many other life patterns are adapted to moisture cycles (Hidore 1974; Fig. 4.6).

# Fire Regimes

In the past, forest fires occurred at different magnitudes and frequencies in different climatic-vegetation types. In the boreal forest, for example, infrequent large-magnitude fires carried the flames in the canopy of the vegetation (crown fires), killing most of the forest. Other environments, such as lower-elevation ponderosa pine (*Pinus ponderosa*) forest in the western United States, had a regime of frequent, small-magnitude, surface fires. Here, the burning was restricted to the forest floor, and most

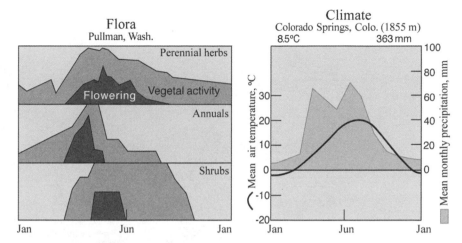

**Figure 4.6.** Annual cycles adapted to temperature and moisture in a climax steppe community. *Left* is from Daubenmire (1968), p. 11. From Daubenmire. *Plant communities: a text book on plant synecology.* © 1968 by Rexford Daubenmire; reprinted with permission from Pearson Education Inc. Figure in the *right* is redrawn from Walter et al. (1975).

mature trees survived. Variations of these two fundamental fire types also occurred.

Precolonial fire regimes for different vegetation types in North America have been determined by analyzing fire scars on living trees. In areas lacking trees, the development of vegetation after recent fires, and early journal accounts and diaries may be used to make an inference about the fire regime.

Precolonial fire regimes in the United States are possible to correlate with climatic-vegetation regions (Vale 1982; Fig. 4.7). A few vegetation types were free from recurrent fires. Tundra, alpine, and warm desert environments had too little fuel for fires. Certain forests of New England, forests in moist topographic situations, and forests in the southern Appalachians apparently were not strongly influenced by fire. Most other forested environments burned with some regularity, although the frequency was highly variable, and both crown and surface fires affected them. Areas with an abundance of herbaceous vegetation seemed to have fire regimes of frequent surface burns.

In some situations, European settlement increased fire frequencies or intensities. Settlers may have increased fire frequency either through carelessness or clearing forests to encourage the growth of grass.

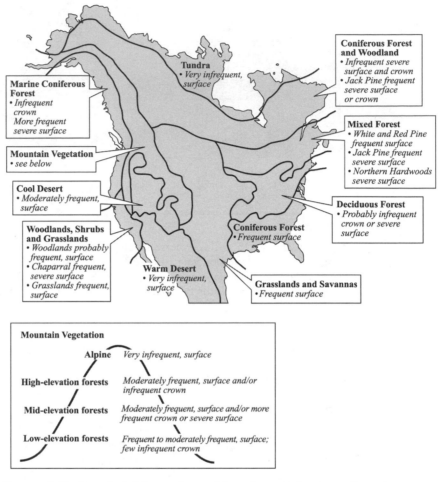

**Figure 4.7.** Precolonial fire regimes of broad vegetation types (based on ecoregions) in North America. Only major divisions of the ecoregion map are shown. From Vale (1982), p. 19; reproduced with permission from Association of American Geographers.

# Plant Productivity

Geographers have found that they can match gross figures of plant productivity with climatic statistics. For example, Bazilevich et al. (1971) found that annual plant production and total plant mass follow a sequence of high values in humid regions and low values in arid regions. These variations are similar to major climate zones. Maximum plant production is associated with the tropics, where combinations of heat and moisture favor maximum production (Table 4.1). Various estimations for potential plant growth from climatic elements have been developed. All give similar global patterns but differ in important detail. Bogorov (1962) reported similar correspondence of productivity with geographic zones of the open oceans.

**Table 4.1.** Distribution of potential production on the earth[a]

| Climate zones | Primary production | |
| --- | --- | --- |
| | Total (t/yr $\times 10^9$) | Average (t/ha/yr) |
| Polar | 1.33 | 1.6 |
| Boreal | 15.2 | 6.5 |
| Temperate | | |
|    Humid | 9.34 | 12.6 |
|    Semiarid | 6.44 | 8.2 |
|    Arid | 1.99 | 2.8 |
| Subtropical | | |
|    Humid | 15.9 | 25.5 |
|    Semiarid | 11.5 | 13.8 |
|    Arid | 7.14 | 7.3 |
| Tropical | | |
|    Humid | 77.3 | 29.2 |
|    Semiarid | 22.6 | 14.1 |
|    Arid | 2.62 | 2.0 |

[a]From Bazilevich et al. (1971).

# Litter and Decomposition

The above ground litterfall forms a litter layer on the ground. The thickness varies which is determined by the rate at which litter is supplied and decomposed. Rates at which litter is added and decomposed vary greatly in different climatic zones, as does the total amount of litter (Table 4.2).

Carbon sequestration has become important due to the link between possible climate change and the accumulation of greenhouse gases in the atmosphere. Coarse and fine woody debris are substantial forest ecosystem carbon stocks. Forest woody detritus production and decay rates depend on climatic conditions. Using forest inventory data, Woodall and Liknes (2008) found that mean forest woody debris carbon vary by climatic regions across the United States. The highest carbon stocks were found in regions with cool summers while the lowest carbon stocks were found in arid desert/steppes or temperate humid regions. Carbon stocks

**Table 4.2.** Decomposition rates of broadleaf and needle litter in selected climates (ecozones)[a]

| Ecozone | $k$ = rate of decomposition (annual litter input/litter accumulation) | $3/k$ = period (in years) until 95% decomposed |
|---|---|---|
| Polar/subpolar: tundra | 0.03 | 100 |
| Boreal | 0.21 | 14 |
| Temperate midlatitudes | 0.77 | 4 |
| Dry midlatitudes: grass steppes | 1.5 | 2 |
| Seasonal tropics | 3.2 | 1 |
| Humid | 6.0 | 0.5 |

[a]From Swift et al. (1979), in Schlutz (1995).

were found to be positively correlated with available moisture and negatively correlated with maximum temperature.

# Controls over the Climatic Effect and Scale

The factors that control the climatic effect change with scale. We can distinguish climatic differences and their controls on different levels or scales. For example, we can detect air temperature differences over a distance of 10,000 km on a global level (related to latitude) and over a few hundred meters in mountain areas (related to exposure or aspect) (Fig. 4.8).

Understanding these climatic controlling factors on a scale-related basis is key to setting ecosystem boundaries.

Notwithstanding the difficulties with using vegetation to delineate ecosystems (cf. Chapter 3), macrofeatures of the vegetation appear to be the appropriate criteria for defining secondary divisions beyond climate (Küchler 1973; Damman 1979). Although only a result, vegetation is important as a criterion in the delineation of geographic zones because it affords a delicate index of climate. The predominance of vegetation also ensures its consideration in any zoning scheme. Usually, the boundaries of vegetationally defined regions coincide with landform units of major relief. This strengthens the primary division. However, the surface features are more useful at lower levels of the hierarchy for subdividing the ecoclimatically circumscribed areas.

The concept of climate as expressed by vegetation has been used frequently as the basis for delineating broad-scale ecological regions.

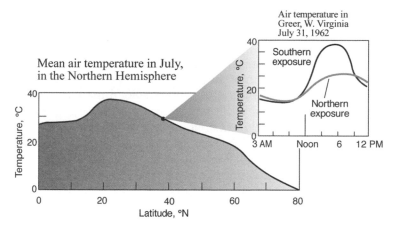

**Figure 4.8.** Temperature at different scales. Figure in the *left* is modified from Isachenko (1973), p. 44; reproduced with permission of John S. Massey (ed.). Figure in the *right* is modified from Smith (1977), p. 150. Copyright © 1977 by Robert Leo Smith; reprinted with permission from HarperCollins Publishers, Inc.

The work of Dokuchaev (1899) was immensely significant in developing the concept. He pointed out that natural conditions are characterized by many common features within the limits of extensive areas (zones) and that these features change markedly in passing from one zone to another. In subsequent studies, Grigor'yev and Budyko (Grigor'yev 1961; Budyko 1974) established that climatic factors determine the boundaries of geographic zones to a considerable extent. Every feature with a distribution that broadly conforms to climate is termed *zonal*. The term *azonal* describes processes or features that occur in several zones. For example, wetlands are not associated with a particular climatic zone.

Efforts to divide the world into ecological regions have been based primarily on the distribution of climate-vegetation zones (e.g., Herbertson 1905; James 1959; Biasutti 1962; Dasmann 1972; Udvardy 1975). Recently, Walter (1977, 1984; Walter and Box 1976) presented a scheme for classifying the world into a hierarchy of ecosystems from a climatic viewpoint. In Russia, Berg (Isachenko 1973) detailed landscape zones based on climate, whereas similar work was developed by Passage (Troll 1971) in Germany and Galoux (Delvaux and Galoux 1962) in Belgium. Some systems for the classification of climates (Köppen 1931; Thornthwaite 1931, 1948) seek to define climatic units that will correspond to major vegetation units. Several authors (e.g., Merriam 1898; Hopkins 1938) have sought to define life zones primarily on the basis of climate. The system of Holdridge (1947; Tosi 1964) uses a complex classification of zones by both temperature and moisture conditions.

In Canada, the concept of forest ecosystem regions (called *site regions*) was developed by Hills (1960a) based on macroclimate. Similar work has been done in other parts of Canada (Crowley 1967; Burger 1976). Krajina (1965) has delineated the biogeoclimatic zones of British Columbia. Climatic regionalization is used in biophysical or ecological land classification throughout Canada (Ecoregions Working Group 1989).

Of the variety of classifications available, the one devised by Crowley (1967) has been adopted as most suitable and is presented in this book to illustrate the basis for regional delineation. A specific application of this classification has been developed and applied to the United States (Bailey 1976, 1995), with later expansion to include the rest of the continents also (Bailey 1989). The system consists of a method for defining successively smaller ecoclimatic regions within larger regions. At each successive level, a different aspect of the climate and vegetation is assigned prime importance in placing map boundaries.

We describe in the following chapters the factors that are thought to differentiate ecoclimatic units and the scale at which they operate.

# CHAPTER 5

# Macroscale: Macroclimatic Differentiation (Ecoregions)

We describe climate primarily in terms of temperature, movement, and water content of (and precipitation from) air masses. Our unique position in the solar system (and possibly throughout much of the universe) combined with our atmosphere enables water to exist on earth in all three states of liquid, solid, and gas. The water molecule is unusual in that its solid form is less dense than its liquid form. This enables ice to float, thus forming huge masses at the poles that strongly affect climate. Climate is more heavily modified by the large land and water masses, the continents, and the oceans. Continental position and landforms, such as mountain chains, also modify climate.

At first, it seems difficult to assess or describe climatic conditions that prevail over large areas, or at the macroscale, because climate changes within short distances due to modification by landform and vegetation. We must, therefore, postulate a climate that lies just beyond those local modifying irregularities. To this climate we apply the term *macroclimate*.

The difference between macroclimate and microclimate is illustrated by the measurements made by Wolfe et al. (1949) in attempting to determine the relationship between climate and various plant communities in the Neotoma Valley in Ohio. They measured the range of variation of climatic variables for the year 1942, first for 88 normal observation stations in representative locations throughout the state (area, 113,000 km$^2$) and then for 109 microclimatic stations in the deep Neotoma Valley, over an area of 0.6 km$^2$. The results are shown in Table 5.1. The greatly varied conditions of microclimate contrast with the comparatively uniform, general climate of Ohio.

R.G. Bailey, *Ecosystem Geography*, DOI 10.1007/978-0-387-89516-1_5,
© Springer Science+Business Media, LLC 2009

**Table 5.1.** Comparison of macroclimate and microclimate[a]

| Meteorological variables | Macroclimate (88 meteorological stations in Ohio) | Microclimate (109 microclimate stations in the Neotoma Valley Ohio) |
|---|---|---|
| Highest annual temperature, °C | 33–39 | 24–45 |
| Time of occurrence of the highest annual temperature | 17–19 July | 25 Apr.–19 Sep. |
| Lowest January temperature, °C | –21 to –29 | –10 to –32 |
| Latest spring frost | 11 Apr.–11 May | 9 Mar.–24 May |
| Earliest autumn frost | 25 Sep.–28 Oct. | 25 Sep.–29 Nov. |
| Days of frost-free period | 138–197 | 124–276 |

[a]From Wolfe et al. (1949); Geiger (1965).

# Causes of Ecoregion Pattern

Variations in macroclimate (as determined by the observations of meteorological stations) are related to several factors: latitude, continental position, and land elevation.

# Latitude

Ecosystem differences on the earth are the result of two primary energy sources. The first is external energy from solar radiation (Fig. 5.1). The primary control of climate at the global level (macroclimate) is variation in solar energy, which is related to latitude. The amount of solar radiation generally decreases from the equator to poles, partly due to increases in the angle of incidence of the sun's rays and partly due to the variation in effective thickness of the atmosphere (Fig. 5.2). This results in generally east–west trending belts or *zones* corresponding to life zones, plant formations, and biomes (Whittaker 1975). Thermal and moisture limits for plant growth determine zone boundaries.

## Thermally Defined Zones

We can delineate three major thermally defined zones (Fig. 5.3): (1) a winterless climate of low latitude, (2) a temperate climate of midlatitudes with both a summer and winter, and (3) a summerless climate of high latitude. In winterless climate, no month of the year has a mean monthly temperature lower than 18°C. The 18°C isotherm approximates

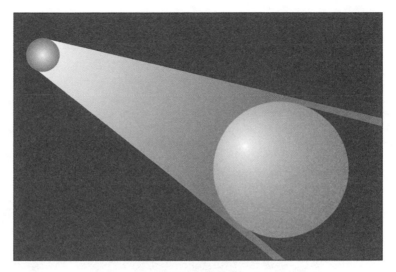

**Figure 5.1.** External (solar) energy source.

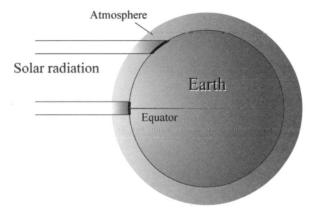

**Figure 5.2.** Solar radiation decreases with latitude because the rays are spread over a larger area and because they pass through a thicker layer of reflecting and absorbing atmosphere.

the position of the boundary of the poleward limit of plants characteristic of the humid tropics. In summerless climate, no month has a mean monthly temperature higher than 10°C. The 10°C isotherm closely coincides with the northernmost limit of tree growth; hence, it separates the regions of boreal forest from the treeless tundra (Fig. 5.4).

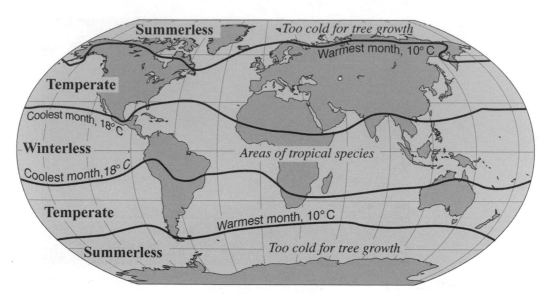

**Figure 5.3.** Zones determined by thermal limits. From Strahler (1965), p. 103; reproduced with permission of the author.

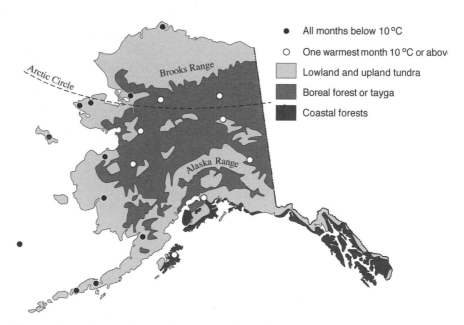

**Figure 5.4.** The northern and western edges of the boreal forest (tayga) in Alaska correspond closely to a line beyond which all months are below 10°C. Climate data from Walter and Lieth (1960–1967) and Walter et al. (1975); vegetation from Viereck et al. (1992).

This scheme gives a satisfactory general picture of the zones. The boundaries between the zones, however, are imprecise. We can illustrate this imprecision by looking more closely at one of the major zones, the boreal coniferous forest. Here the boundary is not marked by an abrupt discontinuity, but by a continuous gradient. Generalizing, we can say that the boreal forest zone is a climatically determined ecological unit covered by a conifer-dominated forest. The Arctic tree line convention-ally marks its poleward limit. In practice, however, we see not an abrupt line but an interpenetrating belt where tree growth is confined to the most favorable sites. Muskeg and bog occupy the wetter sites, with tundra on the exposed ridges (Fig. 5.5).

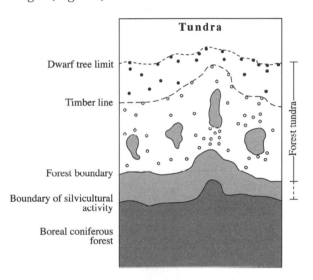

**Figure 5.5.** The boundary between boreal coniferous forest and tundra is usually a transition zone rather than a sharp line. From Hustich (1953), p. 150; reproduced with permission.

The relative amplitudes of annual and diurnal energy cycles vary in each region (Fig. 5.6). Within the tropics, the diurnal range is greater than the annual range. Within temperate zones, the annual range exceeds the diurnal range, although the diurnal can be very large. Within the polar zones, the annual range is far greater than the diurnal range.

## Moisture-Defined Zones

Precipitation and runoff also follow a zonal pattern, generally decreasing with latitude (Fig. 5.7). Near the equator is a zone experiencing conver-gence of air masses. The trade winds moving toward the equator pick up

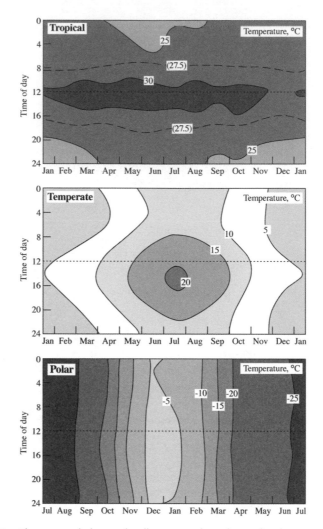

**Figure 5.6.** Thermoisopleth graphs illustrating the relationship between the annual and diurnal energy cycles in tropical, temperate, and polar zones. Stations are Singapore, Oxford, and McMurdo Sound, respectively. From Troll (1966).

moisture over the oceans and, when lifted in the equatorial convergence zone, yield abundant precipitation. Subtropical high-pressure cells centered on the tropics of Cancer and Capricorn (23.5°N and S) control belts of lower rainfall (Fig. 5.8). These zones are too dry for tree growth. Thus, the boundary between the boreal zone and the mid-latitude grasslands in Siberia and the Canadian prairie areas is controlled by the dryness of the climate rather than by its temperature.

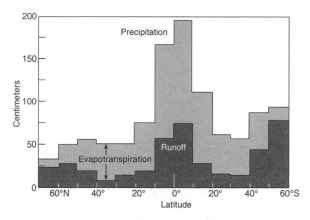

**Figure 5.7.** Distribution of annual precipitation and runoff amounts averaged by latitudinal zones. The vertical difference between the two lines represents the loss through evapotranspiration. From L'vovich and Drozdov in Trewartha et al. (1967), p. 413. Copyright © 1967 by McGraw-Hill, Inc.; reproduced with permission from McGraw-Hill, Inc.

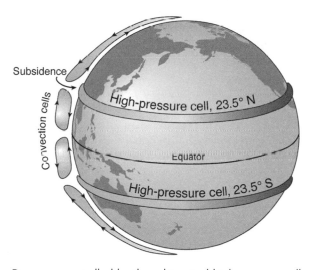

**Figure 5.8.** Dry zone controlled by the subtropical high-pressure cell.

# Continental Position

On a macroscale, continental position also controls climate patterns. If the earth were all land or all water the macroclimates would form simple latitudinal zones, but differences between climates over the land and water modify this arrangement (Fig. 5.9). Because of the greater thermal

**Figure 5.9.** Solar energy modified by the presence of continents.

capacity of water, land heats and cools much more rapidly in response to solar radiation. The difference is great enough to measure not only a diurnal effect but also an annual effect. Because the heat radiating from the land and water affects the surrounding air temperature, air over large bodies of water will have less variable temperatures than air over the land. The land will heat up by day, in the summer, more rapidly than water. It will cool more rapidly in winter. At a given latitude, the summers are warmer and the winters colder on land than on the seas. Warmest months increase with increasing latitude and distance from the sea. This forms a distinction between *marine* and *continental* climates (Fig. 5.10).

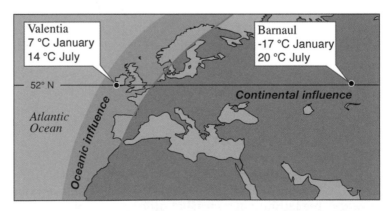

**Figure 5.10.** Marine and continental climates. From Bunnett (1968), p. 119; reprinted with permission from Pearson Education Ltd.

This contrast is reflected in the arrangement of temperatures. Figure 5.11 shows the generalized position of two lines of equal temperature (isotherms) crossing a generalized continent of low elevation. In the Köppen system used in this book (see p. 62, and Table 6.1, p. 81), two isotherms are of special importance. These are the isotherms for the average temperature of the coldest month of –3°C and the isotherm for the average temperature of the warmest month of 22°C. Because the land is colder than the water in winter, the coldest month isotherm bends equatorward as it crosses the continent. Because the land is warmer than the water in summer, the summer isotherm bends poleward in crossing the continent.

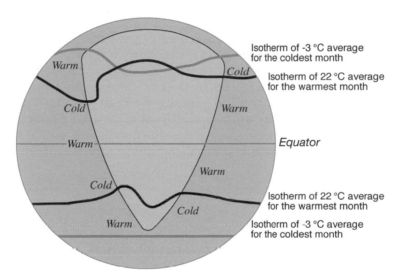

**Figure 5.11.** Generalized global pattern of summer and winter isotherms. From James (1959), p. 181. Copyright © 1959 by Ginn and Company; reprinted with permission from John Wiley & Sons, Inc.

The two isotherms in the Northern Hemisphere, however, are not arranged symmetrically on the continents. The summer isotherm bends sharply equatorward because of the very cold water along the west coast between 35° and 15° N latitude. The winter isotherm is much farther north on the west coast because warm water bathes the west coast beyond 40° latitude and cold water bathes the east coast beyond 40° N.

The position of these isotherms in the Northern Hemisphere reflects another control of climate—the prevailing surface winds. The western sides of the continents poleward of 30° receive westerly winds. The moderating effect of the oceans is felt farther inland on the western sides of the continents. The east-coast climates are generally more continental than those of the west coast at the same latitude. This "stretches" the distance between isotherms so that annual temperature difference between, for example, northern Norway and southern Morocco, which are 2500 km apart, is about the same as that between Newfoundland and Florida, which have only half the distance separating them.

The distribution of land and sea also complicates precipitation patterns. Rain is produced through the cooling of moist air (i.e., air containing large amounts of water vapor). Air picks up water vapor by evaporation. Because evaporation is much more rapid from warm water than from cold water, the warm ocean water is the major source of atmospheric moisture. Figure 5.12 shows that the oceans are symmetrically marked off into regions of warm and cold water, as well as water of intermediate temperature. Rainfall is generally greater over the warm water and over margins of the continents bathed by warm water. Rainfall is small over cold water or over the continental margins bordered by cold water. As can be seen from Figure 5.13, the dry zone (too dry for trees), which is controlled by the subtropical high-pressure cell (see Fig. 5.8), is shifted to the west side of the continents adjacent to these cold waters. The dry zones would extend completely across the continents, except for warm water on the eastern side and associated increased rainfall. We must understand the

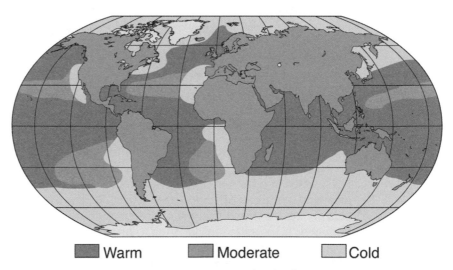

**Figure 5.12.** Ocean temperatures. From Gerhard Schott in James (1959), p. 632. Copyright © 1959 by Ginn and Company; reprinted with permission from John Wiley & Sons, Inc.

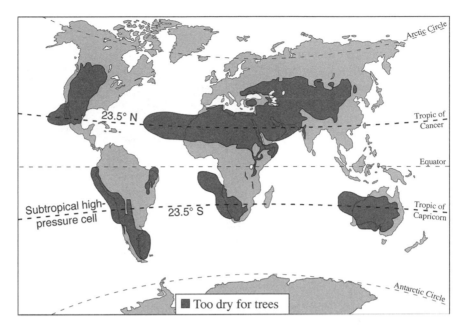

**Figure 5.13.** Zones determined by moisture limits.

oceans, therefore, to understand the continents, because they control the continental climate patterns.

## Climatic Subzones

By combining the thermally defined zones with the moisture zones, we can delineate four ecoclimatic zones: humid tropical, humid temperate, polar, and dry (Fig. 5.14). They are arranged in a regular repeated pattern with reference to latitude and continental position. Within each of these zones, one or several climatic gradients may affect the potential distribution of the dominant vegetation. Within the humid tropical zone, for example, we can distinguish rainforests that have year-round precipitation from savannas that receive seasonal precipitation (Fig. 5.15). Thus, we can subdivide the humid tropical zone, based on moisture distribution (Fig. 5.16) into *climatic subzones*. We can subdivide the other zones similarly.

Locating the boundaries of broad-scale ecosystems requires taking into account visible and tangible expressions of climate such as vegetation. Generally each climate is associated with a single plant forma-

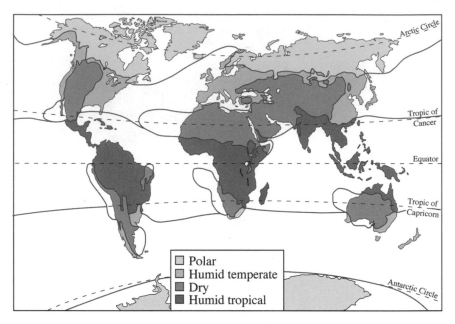

**Figure 5.14.** Four ecoclimatic zones of the earth.

tion (cf. Box 1981, Table 5.2), and is characterized by a broad uniformity both in appearance and in composition of the dominant plant species. Usually a significant correspondence with soils occurs because climate also strongly dominates soil-forming processes.

## Köppen Climate Classification System

The analysis of each zone identifies several climatic subzones. These climatic subzones are correlated with actual climatic types, using the system of climatic classification developed by Wladimir Köppen (1931) and modified by Trewartha (1968). Köppen's system is simple, based on quantitative criteria, and correlated well with the distribution of many natural phenomena, such as vegetation and soil.

Other bioclimatic methods for mapping zones at global levels exist (e.g., Thornthwaite 1931; Holdridge 1947; Troll 1964; Walter et al. 1975). All use selected climatic characteristics that outline zones within which certain general-level-vegetation homogeneity should be found. They also suggest a strong similarity of vegetation in equivalent bioclimatic zones

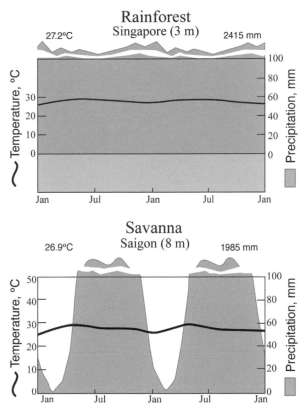

**Figure 5.15.** Rainforest and savanna precipitation patterns. Both environments have a year-round, high-energy input. From Walter et al. (1975).

in different parts of the globe. All the methods appear to work better in some areas than in others and have gained their own following. Köppen's system has become the most widely used climatic classification for geographic purposes. It has become the international standard and is presented in this book to illustrate the basis for zone delineation. Furthermore, among the existing climate classification systems, the one by Köppen (which is primarily based on precipitation and temperature) is found to be the least demanding on data. As meteorological stations around the world routinely collect values of these attributes and the information is generally available in existing maps, this was seen as an additional advantage, as it draws on a relatively consistent global distribution of input data. Other global climate classification systems (for example, Thornthwaite and Holdridge) call for evapo-transpiration data, which is not uniformly available at the global level. A review of the

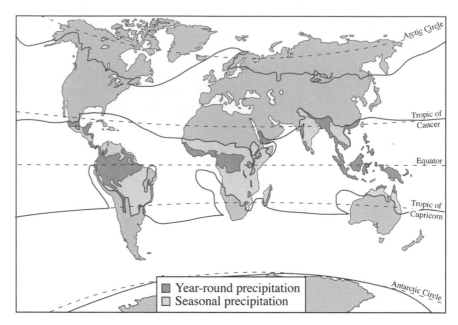

**Figure 5.16.** Subdivision of the humid tropical zone based on moisture distribution.

**Table 5.2.** Broad plant formations and groups of climates

| Formation | Köppen climate group |
| --- | --- |
| Tropical rain forest | A (tropical rainy climates) |
| Tropical desert | B (dry climates) |
| Temperate deciduous forest | C (warm temperate climate) |
| Boreal forest | D (snowy-forest climates) |
| Tundra | E (polar climates) |

Holdridge system for classifying world climatic zones (called life zones) in relation to ecosystem mapping is presented by Lugo et al. (1999).

Others have followed this precedent for using Köppen–Trewartha in global ecological zoning. One good example was carried out by the Foreign Agricultural Organization (FAO) of the United Nation, who used Köppen–Trewartha in combination with vegetation characteristics for development of their ecological zone map for the Forest Resource Assessment 2000 (FAO 2001).

The following 13 basic climates result from applying Köppen's system:

| | |
|---|---|
| Ar | Tropical wet: all months above 18°C and no dry season |
| Aw | Tropical wet-dry: same as Ar but 2 months dry in winter |
| BSh | Tropical/subtropical semiarid: evaporation exceeds precipitation, and all months above 0°C |
| BWh | Tropical/subtropical arid: one-half the precipitation of BSh and all months above 0°C |
| BSk | Temperate semiarid: same as BSh but with at least 1 month below 0°C |
| BWk | Temperate arid: same as BWh, but with at least 1 month below 0°C |
| Cs | Subtropical dry summer (Mediterranean): 8 months 10°C, coldest month below 18°C, and summer dry |
| Cf | Subtropical humid: same as Cs but no dry season |
| Do | Temperate oceanic: 4–7 months above 10°C, coldest month above 0°C |
| Dc | Temperate continental: same as Do but with coldest month below 0°C |
| E | Boreal or subarctic: up to 3 months above 10°C |
| Ft | Tundra: all months below 10°C |
| Fi | Polar ice cap: all months below 0°C |

The distribution of these climates is shown in Figure 5.17. A particular type of climatic regime clearly defines each climatic subzone. Chapter 7 will include diagrams of climates thought to be typical of the 13 climates within the United States.

With few exceptions, the subzones largely correspond to zonal soil types (Bockheim 2005) and zonal vegetation. These zones seem to conform to the concept of "ecocomplex" proposed by Polunin and Worthington (1990), to the extent that they are composed of spatially related ecosystems. Table 5.3 shows the relations between zonal types and climates as classified by Köppen.

Zonal soil types and vegetation occur on sites supporting *climatic climax vegetation*. Such sites are uplands (i.e., sites with well-drained surface, moderate surface slope, and well-developed soils). The climax vegetation corresponds to the major plant formation (e.g., deciduous forest) that is the presumed result of plant succession, given enough time. The system presented here follows the traditional view of succession and follows the polyclimax concept (Tansley 1935). Climatic climax ecosystems reflect the primary influence of the regional climate. Other types of climax ecosystems include edaphic climaxes, among others.

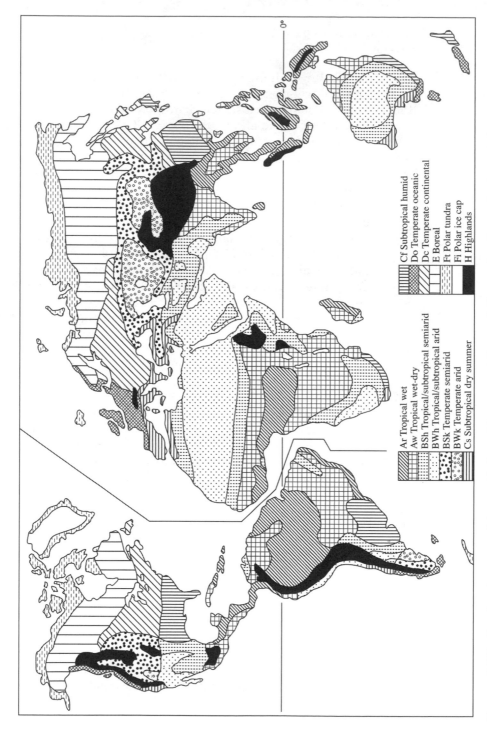

**Figure 5.17.** Ecoclimatic zones of the world. From "Groups and Types of Climates" in Trewartha (1968), frontispiece. From Trewartha. *An introduction to climate.* 4E. © 1968 by McGraw-Hill, Inc.; reproduced with permission of the McGraw-Hill, Companies.

Ar Tropical wet
Aw Tropical wet-dry
BSh Tropical/subtropical semiarid
BWh Tropical/subtropical arid
BSk Temperate semiarid
BWk Temperate arid
Cs Subtropical dry summer

Cf Subtropical humid
Do Temperate oceanic
Dc Temperate continental
E Boreal
Ft Polar tundra
Fi Polar ice cap
H Highlands

**Table 5.3.** Zonal relationships among climate, soil, and vegetation[a]

| Eco-climatic subzone | Zonal soil type[b] | Zonal vegetation |
|---|---|---|
| Ar | Latisols (Oxisols) | Evergreen tropical rainforest (selva) |
| Aw | Latisols (Oxisols) | Tropical deciduous forest or savanna |
| BS | Chestnut, Brown soils, and Sierozem (Mollisols, Aridisols) | Shortgrass |
| BW | Desert (Aridisols) | Shrubs or sparse grasses |
| Cs | Mediterranean brown earths | Sclerophyllous woodlands |
| Cf | Red and Yellow Podzolic (Ultisols) | Coniferous and mixed coniferous-deciduous forest |
| Do | Brown Forest and Gray-Brown Podzolic (Alfisols) | Coniferous forest |
| Dc | Gray-Brown Podzolic (Alfisols) | Deciduous and mixed coniferous-deciduous forest |
| E | Podzolic (Spodosols and associated Histosols) | Boreal coniferous forest (tayga) |
| Ft | Tundra humus soils with solifluction (Entisols, Inceptisols, and associated Histosols) | Tundra vegetation (treeless) |
| Fi | | |

[a]From Walter (1984, p. 3).
[b]Names in parenthesis are soil taxomony soil orders (USDA Soil Conservation Service 1975).

## Subdivisions of a Subzone

Fine-scale climatic variation can be used to delineate smaller ecological regions. Climate is not completely uniform within climatic subzones as local contrasts break up and differentiate the major subcontinental zones, so further subdivision can be undertaken. Within the arid subzone, for example, deserts that receive only winter rain (such as the Sonoran Desert) can be distinguished from those that receive only summer rain (such as the Chihuahuan Desert). Similarly, the vegetation of the savanna subzone is highly differentiated (Fig. 5.18): it has heavy forest near its boundary with the equatorial zone and sparse shrubs and grasses near its arid border. Variation in the length and intensity of the rainy season relates to both the variety of vegetation and to soil and hydrologic conditions; Canadian geographer Kenneth Hare (1950) recognized three subzones within the boreal zone of the northern hemisphere: closed-crown forest, woodland, and forest-tundra.

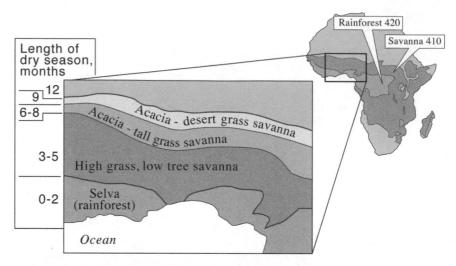

**Figure 5.18.** Subdivision of the savannas of central Niger. From Shantz and Marbut in James (1959), p. 304. Copyright © 1959 by Ginn and Company; reprinted with permission from John Wiley & Sons, Inc.

# Elevation

Lastly, macroclimate is related to elevation. Up to this point, we have been discussing the simple climatic patterns as they would develop on a flat continent in which climate is a function of latitude and continental position (Fig. 5.19). Internal energy from radioactivity and primordial heat (Fig. 5.20) causes mantle convection, plate tectonics, and mountain building. This process creates patterns of high mountains on the continents, which further modify and distort the climatic pattern created by solar energy and the difference between water and land. These mountains are arranged without conforming at all to orderly latitude zones of climate. They cut across them irregularly. For example, we find volcanic mountains in the cold deserts of Antarctica as well as near the equator in Central America.

An essential feature of mountainous regions is vertical differentiation of climate and vegetation based on the effects of elevation change. Land above sea level ranges up to 8.9 km (Mt. Everest in the Asian Himalayas) (Fig. 5.21). The direct effect of these elevational changes on the characteristic of lowland and highland environments is striking. At an altitude of 8 km, the density of the atmosphere is less than one-half its density at sea level. With this thinner shell of atmosphere above them, high elevations receive considerably more direct solar radiation than sea-level locations. Within the lower layers of the atmosphere, temperature decreases

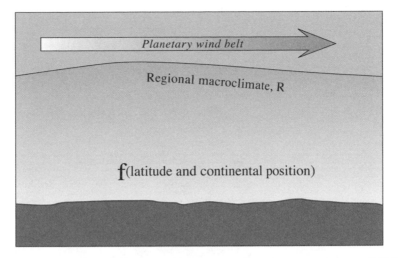

**Figure 5.19.** Scales of climatic zones: macroclimates. From Yoshino (1975) in Barry (1992), p. 12; reprinted with permission from Routledge.

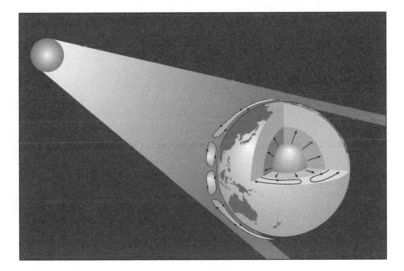

**Figure 5.20.** Internal energy source.

with elevation. Average temperatures drop by about 6.4°C per 1000 m. However, this rate of decrease is not uniform. Temperatures at any given altitude decrease from the equator toward the poles (Fig. 5.22). Highlands extending into the atmosphere encounter colder temperatures, depending on latitude.

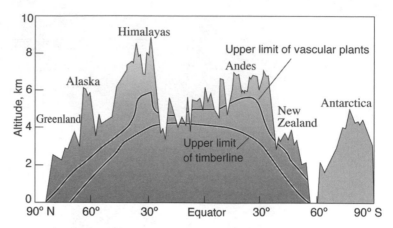

**Figure 5.21.** Elevation and plant growth. Changes in the elevational levels of plant growth in mountain areas at different latitudes. From Swan (1968); reprinted with permission from Indiana University Press.

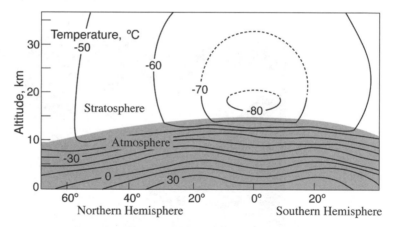

**Figure 5.22.** Vertical arrangement of average annual isotherms. From Clayton and Ramanathan in James (1959), p. 423. Copyright © 1959 by Ginn and Company; reprinted with permission from John Wiley & Sons, Inc.

The effects of elevation on temperature are twofold: the average temperature decreases and the daily range increases with higher elevation. Both effects are due to the clearer, more rarified air that allows more solar radiation to reach the ground surface (raising the midday temperature) and also permits more rapid heat radiation from the ground at night. The net effect is to replicate over short vertical distances the temperature changes we encounter in latitudinal change over considerably greater distances.

The elevation limits of the various types of vegetation also correspond to vertical temperature distribution. Roughly the same succession of vegetation types is found on a mountain, according to the location of its zone (see Fig. 5.21). The vertical differentiation of vegetation and other forms of life reaches a maximum in low latitudes. Here we find the greatest variety of elevational zones.

We can detect the effect of elevation on plant cover within smaller areas, as well as on a world scale. In the Northern Hemisphere, for example, a south-facing slope receives more insolation than a flat surface, and a north-facing slope receives less. Thus the same temperature conditions found on a tableland may occur at a higher altitude on a nearby south-facing slope and at a lower altitude on a north slope. The distribution of vegetation is correlated with temperature and the resulting moisture differences. Because of this, a particular plant community will be found above its ordinary altitudinal range on south slopes and below it on north slopes. In Arizona, ponderosa pine (*Pinus ponderosa*) from the montane zone may come down to 1800 m on north-facing slopes, whereas pinyon-juniper from the lower woodland zone may be found extending upward to 2100 m on south-facing slopes (Fig. 5.23). In general, a vegetation zone extends higher on the south side of a mountain than on the north side. Note that the effect of elevation is modified by the direction of moisture-bearing air and by variations in cloud cover with elevation. This is particularly true for small tropical islands where rainforest covers mountain slopes facing the trade winds and dry desert scrub, dry forest, or savanna covers the leeward side.

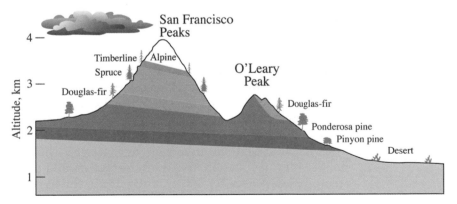

**Figure 5.23.** Vegetation zonation on San Francisco Peaks, Arizona, as viewed from the southeast, illustrating the effects of northern and southern exposures. From Merriam (1890).

Other climatic elements reinforce the effects of temperature on vertical differentiation. Rainfall, for instance, develops a somewhat vertical zoning. Up to 2 or 3 km of elevation in the middle latitudes, for example, the rainfall on mountain slopes increases. Unbroken ranges of mountains are effective barriers to the passage of moisture (Fig. 5.24). The mountain ranges along the Pacific Coast of the United States, for example, intercept moisture transported from the Pacific Ocean by prevailing westerly winds, so that coastal ranges are moist and inland regions are dry.

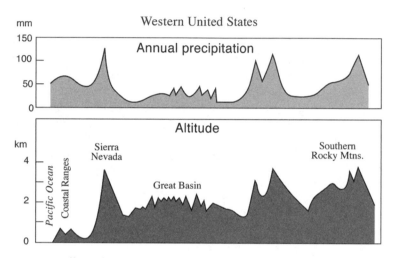

**Figure 5.24.** Effect of altitude on precipitation across the western United States at approximately 38° N. From Bailey (1941), p. 192.

## Effect of Latitude on Elevational Zonation

Just because a place at considerable elevation in the low latitudes has the same average temperature as a place at sea level in middle or high latitudes, we cannot assume it also has the same climate. The climate of an elevational belt in the mountains is always different from the climate of a northerly climate zone. A coniferous forest in the montane zone of a low-latitude mountain is not the high-elevation equivalent of a coniferous forest at high latitude, such as the tayga. Differences include day length, solar declination, length of the season, and precipitation pattern. Precipitation generally increases upward in the mountains as a result of ascending air masses.

High elevation produces variations of lowland climates. Such high-elevation areas do not have the same climate as the adjacent lowland, but they do have the same *climatic regime* (cycle of weather phenomena)

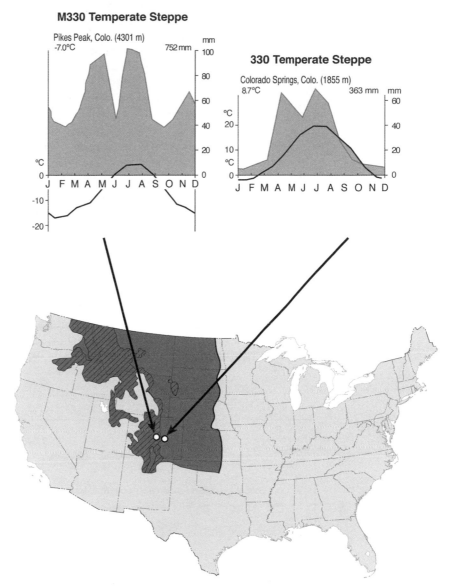

**M330 Temperate Steppe**

Pikes Peak, Colo. (4301 m)
-7.0°C                    752 mm

**330 Temperate Steppe**

Colorado Springs, Colo. (1855 m)
8.7°C                    363 mm

**Figure 5.25.** Climatic comparison of a mountain range embedded in a zone: the Rocky Mountains, a temperate steppe regime highlands. Data from Walter et al. (1975).

(Fig. 5.25). These systems do not belong to the lowland climate zones; rather, each mountain range forms a certain ecological unit in itself, whose character depends on the climatic zone(s) where it is located.

This results in a certain zonal relationship, which also affects the vertical climate-vegetation-soil zonation within the mountain range.

Between the individual elevational belts, a lively exchange of materials occurs: water and the products of erosion move down the mountains; updrafts and downdrafts carry dust and organic matter; animals move easily from one belt into the next; and wind and birds spread seeds. The elevational belts, as a result, are not always as sharply separated from each other as are the climatic zones. The geographic area over which a sequence of belts extends is considered to be a large ecological unit. In this sense, we do not treat the montane forest belt as a separate climatic zone. The montane belt is only one member of the total sequence of altitudinal belts. Montane belts in mountainous areas of different climatic zones are just as distinct from one another as the montane belt is from other altitudinal belts in the same zone. For example, the montane coniferous belt appears in the subarctic zone as spruce-fir forest, whereas fir and pine forest represent it in the steppe zone.

If a mountain range occurs in several climatic zones, it produces different vertical zonation patterns in each zone. The number of elevational zones (or belts) present and their relative elevational positions depend on the lowland climatic zone. This shows in Figure 5.26, which compares locations in the Rocky Mountains. In the temperate steppe (semiarid) climatic portion, the lowermost zone is a sagebrush (*Artemisia*) basal plain;

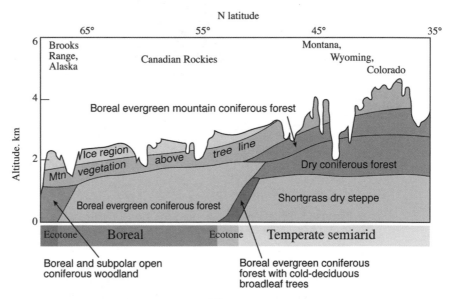

**Figure 5.26.** Vertical zonation in different ecoclimatic zones along the eastern slopes of the Rocky Mountains. From Schmithüsen (1976), p. 70.

this is followed by a montane zone of Douglas-fir (*Pseudotsuga menziesii*) and spruce and fir (*Picea-Abies*); above is the subalpine zone followed by alpine tundra and then perennial ice and snow (Fig. 5.27). This sequence of altitudinal zones repeats on mountain ranges throughout the lowland climatic zone.

**Figure 5.27.** Elevational zonation in the Sawtooth Mountains of Idaho, a high mountain landscape within a temperate steppe climatic regime. Photograph by Frank M. Roadman, Soil Conservation Service.

Soils also change their character with increasing elevation, responding to the changes in climate and vegetation. Figure 5.28 shows how soil profiles change with elevation zones in the western United States.

The effect of latitude on elevational zonation has been recognized for many years going back to Alexander von Humboldt (1769–1859). His travels in the Andes led him to categorize the three-dimensional nature of mountains and to recognize the influence of latitude on the elevation of the distinctive altitudinal belts (Ives et al. 1997). Much later, Carl Troll developed a classification of the mountainous regions of the tropical Americas (1968) and then expanded it worldwide (1972). The latitudinal variation in southern Rocky Mountain forests was analyzed by Peet (1978).

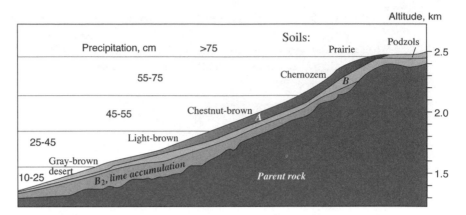

**Figure 5.28.** Gradation of soils from a dry steppe-climate basin (*left*) to a cool, humid climate (*right*) ascending the west slope of the Big Horn Mountains, Wyoming. (Note that the soil profile is an extreme exaggeration for the purpose of illustration.) From Thorp (1931).

## Effect of Latitude and Elevation on Inland Waters

Elevation and latitude also affect inland waters. According to Illies (1974), specialized benthic communities live in the upper reaches of rivers (*rhithron*) where high oxygen concentrations, low temperature fluctuations, and strong currents prevail. These communities show extremely well those morphological and physiological specializations that are adaptive responses to the flowing milieu. They include mayfly and stonefly larvae, freshwater limpets, amphipods, some caddis larvae, and some chironomid larvae. Winter spawners predominate among the fish (trout, char, grayling). In the lower reaches (*potamon*), temperature fluctuations are greater, and oxygen concentrations are more variable and lower. In addition, the current near the bottom is usually weak. Eury-thermous species, tolerant to oxygen deficits, live here and, for the most part, are to be found in standing waters. They include dragonfly lar-vae, water beetles, warm-adapted caddis flies, water lice, and numerous water mites. The fish always spawn in summer (carp, pike, and perchlike forms). The communities of the rhithron and potamon regularly occur throughout the world in relation to latitude and elevation (Fig. 5.29). The only exceptions are locations where the extension of the rhithron community is impeded, as occurs on islands and isolated mountain chains.

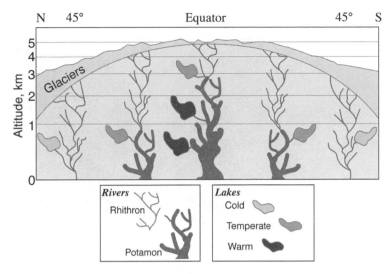

**Figure 5.29.** Elevation and latitude effect on inland waters. From Illies (1974), p. 40; reproduced with permission from Macmillan Press, Ltd.

Lakes show similar relationships. On the basis of their pattern of thermal stratification, we can distinguish three different types (Fig. 5.29). Cold lakes of the polar regions have their warmest water in the lower layers. The upper layers are warmed during only a few months of summer and usually remain cold and ice-covered for many months. Warm lakes of the tropics and subtropics never freeze, and throughout the year the bottom layers are coldest. Lakes in temperate latitudes change during the course of the year between both extremes. The change from a warm summer lake to a cold winter lake takes place in autumn, and the reverse takes place in spring.

## Azonal Highlands

In summary, the mountains create orographically modified macroclimates that exhibit elevational zonation. They are irregularly arranged with reference to latitude and continental position. They represent interruptions of the generalized global pattern of climates. Because they can occur in any zone, they are referred to as *azonal* (Fig. 5.30). Figure 5.31 shows a schematic representation of the modification of the macroclimate in mountainous terrain.

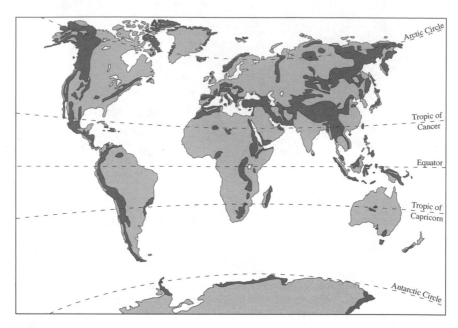

**Figure 5.30.** The arrangement of orographically modified macroclimates on the continents.

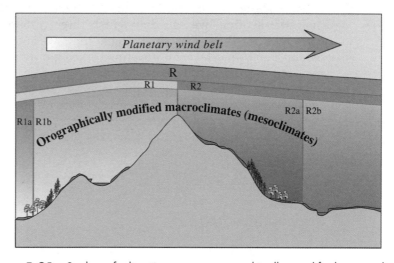

**Figure 5.31.** Scales of climatic zones: orographically modified macroclimates (mesoclimates). From Yoshino (1975) in Barry (1992), p. 12; reprinted with permission from Routledge.

# Macroclimatic Differentiation in Review

The major ecosystems are arranged on the earth's land masses in a predictable pattern. James (1959) provides one the best explanations of this. The pattern involves the interplay of two principles: (1) that all ecosystems causally related to climate are arranged in a regular repeated pattern in relation to latitude and continental position and (2) that all ecosystems causally related to surface features are irregularly arranged with reference to latitude and continental position. If the earth were all land or all water, the world's climates would form simple latitudinal zones. But the differences between the climates over the land and climates over the water modify the simple latitudinal arrangement. The irregular pattern of high mountains on the continents further modifies and distorts the simple climatic patterns that would develop on a flat continent. Therefore, we can describe the actual arrangement of ecosystems by the interplay of the principle of climatic regularity and surface irregularity.

# CHAPTER 6

# Ecoclimatic Zones of the Earth

The effects of latitude, continental position, and elevation, together with other climatic factors, combine to form the world's ecoclimatic zones, herein referred to as an ecosystem region, or ecoregion. Figure 5.17 shows the climatic zones where we might expect distinct ecosystem assemblages to occur. This map shows climatic units that are important to the climatologist and can be used to help determine ecosystem boundaries at the macroscale. Criteria for refining and delineating these zones, or ecoregions, at several levels of detail are presented below.

## Criteria Used in Delineating Ecoregion Levels

Because meteorological stations are too sparse in many areas, data are simply not available to map more precisely the distribution of these ecological climates. Thus, we generally substitute other distributions. Köppen used the composition and distribution of vegetation in his search for significant climatic boundaries, and vegetation is a major criterion in the morphoclimatic maps of Tricart and Cailleux (1972) and the ecosystem region maps of Bailey (1976, 1989; Bailey and Cushwa 1981) and Walter and Box (1976).

Climatic differences useful in recognizing macroscale units can be reflected in the vegetation in several ways (Damman 1979): (1) changes in forest-stand structure, dominant life forms, and topography of organic deposits; (2) changes in dominant species and in the toposequence of plant communities; and (3) displacement of plant communities, changes in the chronosequence of a habitat, and minor changes in the species composition of comparable plant communities. Other differences are given by Küchler (1974) and van der Maarel (1976).

R.G. Bailey, *Ecosystem Geography*, DOI 10.1007/978-0-387-89516-1_6,
© Springer Science+Business Media, LLC 2009

Traditionally, the principal source of such information has been vegetation mapping by ground survey. If large areas are to be surveyed, this approach is not practical. Instead, researchers use satellite remote-sensing data with its synoptic overview to look for zones where vegetation cover is relatively uniform (e.g., Matthews 1983; Soriano and Paruelo 1992). These zones are especially apparent in low-resolution, remote-sensing imagery (Tucker et al. 1985).

In some areas, problems resulting from disturbance and the occurrence of an intricate pattern of secondary succession stages make regional boundary placement difficult. In such areas, we can overcome these problems by considering the patterns displayed on soil maps of broad regions, such as the FAO/UNESCO World Soil Map (FAO/UNESCO 1971–78). Because soils tend to be more stable than vegetation, they provide supplemental basis for recognizing ecosystems, regardless of present land use or existing vegetation. However, they must be used with care because they contain the imprint of past environments (see discussion of "fossil soils" on p. 34).

With these guidelines in mind, we can delineate climate-controlled ecoregions of the world. Three levels can be identified. The broadest are *domains* and within them are *divisions*. They are based mostly on the large ecological climate zones following the 1931 system of Köppen as modified by Trewartha (1968) and summarized in Table 6.1.

**Table 6.1.** Regional climates[a]

| Köppen group and types | Ecoregion equivalents |
| --- | --- |
| A  **Tropical and humid climates** | **Humid tropical domain** (400) |
| Tropical wet (Ar) | Rainforest division (420) |
| Tropical wet-dry (Aw) | Savanna division (410) |
| B  **Dry climates** | **Dry domain** (300) |
| Tropical/subtropical semiarid (BSh) | Tropical/subtropical steppe division (310) |
| Tropical/subtropical arid (BWh) | Tropical/subtropical desert division (320) |
| Temperate semiarid (BSk) | Temperate steppe division (330) |
| Temperate arid (BWk) | Temperate desert division (340) |
| C  **Subtropical climates** | **Humid temperate domain** (200) |
| Subtropical dry summer (Cs) | Mediterranean division (260) |
| Humid subtropical (Cf) | Subtropical division (230) |
| | Prairie (250)[b] |
| D  **Temperate climates** | |
| Temperate oceanic (Do) | Marine division (240) |
| Temperate continental, warm summer | Hot continental division (220) |

**Table 6.1.** (continued)

| Köppen group and types | Ecoregion equivalents |
|---|---|
| (Dca) | Prairie division (250)[b] |
| Temperate continental, cool summer | Warm continental division (210) |
| (Dcb) | Prairie division (250)[b] |
| E **Boreal climates** | **Polar domain** (100) |
| Subarctic (E) | Subarctic division (130) |
| F **Polar climates** | Tundra division (120) |
| Tundra (Ft) | |
| Ice cap (Fi) | |

[a]Based on the Köppen system of classification (1931), as modified by Trewartha (1968) and Trewartha et al. (1967).
[b]Köppen did not recognize the prairie as a distinct climatic type. The ecoregion classification system represents it at the arid sides of the Cf, Dca, and Dcb types.

### Definitions and Boundaries of the Köppen–Trewartha System

| | |
|---|---|
| Ar | All months above 18°C and no dry season |
| Aw | Same as Ar but with 2 months dry in winter |
| BSh | Potential evaporation exceeds precipitation and all months above 0°C |
| BWh | One half the precipitation of BSh and all months above 0°C |
| BSk | Same as BSh but with at least 1 month below 0°C |
| BWk | Same as BWh but with at least 1 month below 0°C |
| Cs | 8 months 10°C, coldest month below 18°C, and summer dry |
| Cf | Same as Cs but no dry season |
| Do | 4–7 months above 10°C and coldest month above 0°C |
| Dca | 4–7 months above 10°C, coldest month below 0°C, and warmest month above 22°C |
| Dcb | Same as Dca but warmest month below 22°C |
| E | Up to 3 months above 10°C |
| Ft | All months below 10°C |
| Fi | All months below 0°C |

A/C boundary = equatorial limits of frost; in marine locations, the isotherm of 18°C for coolest month
C/D boundary = 8 months 10°C
D/E boundary = 4 months 10°C
E/F boundary = 10°C for warmest month
B/A, B/C, B/D, B/E boundary = potential evaporation equals precipitation

# The Domains

## Polar Domain

The polar domain is where frost action primarily determines plant development and soil formation. This domain is divided on the basis of plant development into (1) tundra division, where the average monthly air temperature in all months is below 10°C, and (2) a subarctic tayga division, where the average air temperatures in as many as 3 months of the year may be warmer than 10°C.

## Humid Temperate Domain

The humid temperate domain comprises the humid midlatitude forest of broad-leaved and coniferous trees. The variable importance of winter frost largely determines the divisions: (1) a warm continental division that has very cold winters but warm summers; (2) a hot continental division that has cold winters but hot summers; (3) a subtropical division having mild winters and hot summers; (4) a marine (maritime) division that has mild winters, cool summers, and a minor role played by frost; (5) a prairie transition division that has a subhumid climate; and (6) a Mediterranean-type division with dry summers and mild winters.

## Dry Domain

The dry domain comprises the arid and semiarid regions of middle and adjacent latitudes and has the discontinuous vegetation of steppe, xerophytic bush, and desert types, with only intermittent and local runoff. It is subdivided into divisions according to (1) the amount of water deficit, determining whether it is semiarid steppe or arid desert and (2) the winter temperatures, which influence biological and physical processes and the duration of any snow cover. This temperature factor is the basis of distinction between temperate and tropical/subtropical dry regions.

## Humid Tropical Domain

The humid tropical domain is characterized by persistently high moisture and temperature levels and perennial streamflows. It is divided into (1) the savannas or alternating wet-dry tropics and (2) the rainforest or wet tropics, on the basis of the seasonality of rainfall, total annual rainfall, and density of plant cover.

A generalized diagram of the climatic groups and types (that form the basis for the domains and divisions) as they would exist on a hypothetical continent of low and uniform elevation is shown in Figure 6.1. The diagram shows typical positions of each type and common boundaries between climates.

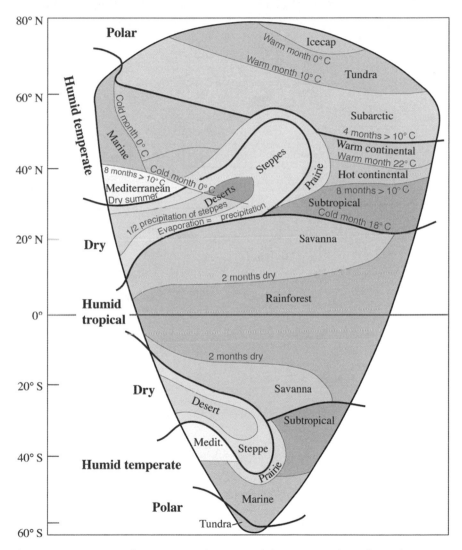

**Figure 6.1.** Pattern of ecoregions (domain and division) on a hypothetical continent of low uniform elevation. Adapted in part from Crowley. Compare with world map, Plate 1.

# The Provinces

Each of the above divisions is further subdivided into *provinces*, on the basis of macrofeatures of the vegetation. These subdivisions express more refined climatic differences than the domains and divisions.

Mountains exhibiting altitudinal zonation and having the climatic regime of the adjacent lowlands are distinguished according to the character of the zonation. Figure 6.2 illustrates the use of these criteria.

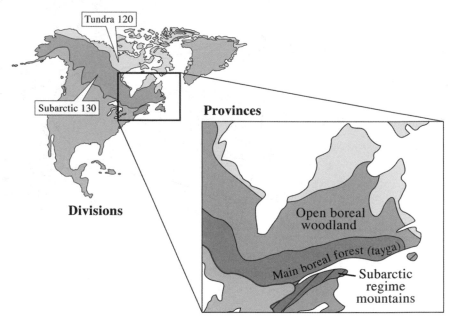

**Figure 6.2.** Variation within an ecoregion domain. Subdivisions of the polar domain in North America.

# Ecoregion Maps

I have applied these criteria to produce maps of the three levels of the ecoregion classification for the United States and the world's continents. A simplified version of the map of the continents, which shows domains and divisions, is reproduced as Plate 1 (inside back cover). This map is based on a world map of natural climate-vegetation landscapes types in the *Fiziko-geograficheskii atlas mira* (Gerasimov 1964). Table 6.2 presents the area contained in each region and its percentage of the global

**Table 6.2.** Approximate area and proportionate extent of ecoregions[a]

|  | km$^2$ | % |
|---|---|---|
| **100 Polar domain** | 38,038,000 | 26.00 |
| 110 Ice cap division | 12,823,000 | 8.77 |
| M110 Ice cap regime mts | 1,346,000 | 0.92 |
| 120 Tundra division | 4,123,000 | 2.82 |
| M120 Tundra regime mts | 1,675,000 | 1.14 |
| 130 Subarctic division | 12,259,000 | 8.38 |
| M130 Subarctic regime mts | 5,812,000 | 3.97 |
| **200 Humid temperate domain** | 22,455,000 | 15.35 |
| 210 Warm continental division | 2,187,000 | 1.49 |
| M210 Warm continental regime mts | 1,135,000 | 0.78 |
| 220 Hot continental division | 1,670,000 | 1.14 |
| M220 Hot continental regime mts | 485,000 | 0.33 |
| 230 Subtropical division | 3,568,000 | 2.44 |
| M230 Subtropical regime mts | 1,543,000 | 1.05 |
| 240 Marine division | 1,347,000 | 0.92 |
| M240 Marine regime mts | 2,194,000 | 1.50 |
| 250 Prairie | 4,419,000 | 3.02 |
| M250 Prairie regime mts | 1,256,000 | 0.88 |
| 260 Mediterranean division | 1,090,000 | 0.75 |
| M260 Mediterranean regime mts | 1,561,000 | 1.07 |
| **300 Dry domain** | 46,806,000 | 32.00 |
| 310 Trop/subtrop steppe division | 9,838,000 | 6.73 |
| M310 Trop/subtrop steppe regime mts | 4,555,000 | 3.11 |
| 320 Trop/subtrop desert division | 17,267,000 | 11.80 |
| M320 Trop/subtrop desert regime mts | 3,199,000 | 2.19 |
| 330 Temperate steppe division | 1,790,000 | 1.22 |
| M330 Temperate steppe regime mts | 1,066,000 | 0.73 |
| 340 Temperate desert division | 5,488,000 | 3.75 |
| M349 Temperate desert regime mts | 613,000 | 0.42 |
| **400 Humid tropical domain** | 38,973,000 | 26.64 |
| 410 Savanna division | 20,641,000 | 14.11 |
| M410 Savanna regime mts | 4,488,000 | 3.07 |
| 420 Rainforest division | 10,403,000 | 7.11 |
| M420 Rainforest regime mts | 3,440,000 | 2.35 |

[a]From World Conservation Monitoring Centre (1992).

land area. In Chapter 7, we discuss the nature of the ecoregion domains and divisions in the United States.

We can also identify ecoregions in the oceans, which together occupy some 70% of the earth's surface (Plate 2, inside back cover). These are determined by the interaction of climate and large-scale ocean currents. The definitions and basis of the map units are presented in the Appendix.

# Ecoregion Boundaries

Ecoregions have relatively smooth boundaries because they are controlled by the macroclimate that *lies above the modifying effects of the earth's surface*. As discussed in Chapter 5, the boundaries actually mark zones of transition, or ecotones, between ecoregions. They are characterized by a mosaic of vegetation types containing interdigitated patches of different types (Fig. 5.5). Mosaic patterns are controlled by patterns in substrate (e.g., soils and microtopography) (Gosz 1993).

The ecoregion boundaries cannot be compiled (or refined) from boundaries of individual types because the latter are not always easily assigned to either of their neighboring ecoregions; certain types are common to both ecoregions (see Bailey 2005).

# Local Contrasts Within Zones

The ecoregions map shows climatically determined ecological units. This *synoptic map* shows mainly global patterns and must not be relied on for local details. When interpreting the map, we should recognize three limitations. First, strong internal variations can occur within regions where (related to elevation, geology, or groundwater) they form a complex intraregional mosaic. Furthermore, slow gradations rather than abrupt discontinuities may mark regional boundaries. A second limitation is that the boundaries of regions are subject to slow but continuous change, related to long-term alterations in climate and, to a lesser extent, to plant succession. Third, the vegetation conditions indicated by the names refer to undisturbed plant cover that is known to exist or is assumed to grow if human interventions were removed. In some regions (e.g., the hot continental), little natural plant cover remains, whereas in others (e.g., the subarctic), the opposite is widely true. Deductions about local situations should not be made without careful study of local literature and preferably on-the-spot inspection.

# Relationship to Other Ecoregional Zoning Systems

A number of approaches for defining ecoregions have grown out of the system presented in this book. For example, The Nature Conservancy (TNC) has shifted the focus from conservation of single species and small sites to conservation planning on an ecoregional basis (The Nature Conservancy 1997). On their map, this text appears:

> "This map was developed as a coordinated effort by TNC US ecoregional planning teams. Based on Bailey 1994, the ecoregions have been modified for both biological and administrative purposes."

They included other criteria, such as the distribution of species as well as the location of conservation units within ecoregions to be administered or managed. Land use may have been included in the mix of factors they considered (Denny Grossman, personal communication).

Hargrove and Luxmoore (1998) created ecoregions of the United States by applying a clustering algorithm to factors derived from climate and soils data. Although the variables were chosen because of their hypothesized relationship to vegetation patterns, this relationship was not formally defined.

A similar approach to ecoregionalization is to overlay thematic maps either manually (Omernik 1987; Gallant et al. 1995; Harding and Winterbourn 1997; Olson et al. 2001) or within a geographic information system (Host et al. 1996; Bernert et al. 1997). Typically, ecoregion boundaries are placed where boundaries of several input layers are located in close proximity to one another. Boundaries defined in this manner represent areas of spatially abrupt changes in ecological characteristics. The overlay method has been referred to as the *weight of evidence* approach (McMahon et al. 2001) and the *gestalt* method (p. 27, this volume; Jepson and Whittaker 2002) because homogeneous-appearing regions are recognized and boundaries are drawn through a process of intuitive and holistic reasoning based largely on visual appearance. Ambiguity arises from the fact that boundaries of input layers rarely conform to one another (the "GIS trap"; see Bailey 1988b). The general procedure is consistent in concept with the empirical approach because it emphasizes pattern over process (Omernik 2004).

The empirical approach seeks to discern patterns in the data. The resultant maps frequently show highly fragmented map units, with small, noncontiguous units of the same region distributed over wide areas. This is particularly true in complex terrain such as the western United States. In the approach presented in this book, the boundaries of terrestrial ecoregions are determined to a considerable extent by the changing nature of the climate over large areas. This approach takes into account compensating factors (see Chapter 11) that override the climatic effect. For example, in the High Plains of the semiarid southwestern United States, forests extend along streams because of the extra water supply. Ponderosa pine and shrub islands within the grasslands of these regions indicate rocky soil conditions, forming reservoirs of water for taproots. These forests occur there because of the ground water condition; not because of the climate. In these cases, a map of climate-based ecoregions ignores such areas and relegates edaphically controlled ecosystems to a lower level of classification and more detailed maps. The same is true for forests that respond to cooler temperatures and additional moisture and are found on north-facing slopes in semiarid mountains (Fig. 1.11). The emphasis in this kind of mapping is on understanding pattern in terms of process.

Geography-based ecosystem mapping does not depict a large number of ecoregion units in mountainous locations. In lowland areas, macroclimate determines the distribution of major ecosystems. Mountains can be distinguished where, as result of elevation, the climate differs substantially from adjacent lowlands to cause complex vertical climate-vegetation-soil zonation. The character of the zonation depends on the climatic zone(s) where the mountain is located. These areas are regionalized by considering the pattern (number, sequence, and elevational position) and composition of the elevational belts or zones. Mountains embedded in similar macroclimates will have similar patterns. Between the individual belts, a lively exchange of materials occurs (Chapter 5). The belts, as a result are interconnected, and the geographic area over which a similar pattern of belts extends is considered to be a large ecological unit—an ecoregion. Like edaphically controlled ecosystems described above, individual belts are not classified and mapped at this level in the hierarchy of ecosystems.

Regarding land use, the ecoregion delineations described in this book are based solely on biophysical factors (i.e., climate, landform, and vegetation). This is an important difference from the systems described above in which land use is one of the delineators. Because land use has not been considered in drawing up regional boundaries, there may not be correspondence between these biophysical-based ecoregions and existing vegetation patterns.

The system in this book also stands in contrast to other systems, which use the range limits of species and races of plants and animals as criteria for determining the boundaries of ecoregions. Sometimes, the range limits for several species might coincide with an ecoregion boundary if that boundary follows some barrier that prevents range expansion, such as where plains meet mountains. Often, however, the range of a species does not stop abruptly at the border of an ecoregion but continues for a distance into the adjacent ecoregion. The reason for this seems to be that some isolated areas of suitable habitat usually occur in the adjacent region. Furthermore, because, at small map scales, physiognomy (lifeform) is the best expression of ecological conditions (Küchler 1973; Gosz and Sharpe 1989), floristic and faunistic differences are best left to maps with other purposes. Because physiognomy is basic and applicable without exception anywhere on Earth, it was selected to serve as the source for the criteria necessary to establish the basis for regional differences. These criteria permit a uniform approach throughout the world and put the various parts of the world on a comparable basis.

Finally, the boundaries of ecoregions on some maps are very irregular; thus implying accuracy and precision. In contrast, the view of ecoregions promulgated in this book looks on the boundaries between ecoregions as representing climatic gradients that reflect gradual change and tend to be indistinct. Such boundaries cannot be located precisely. Complex, squirmy boundary lines on ecoregion maps give a false sense of accuracy and precision.

# CHAPTER 7

# Ecoregions of the United States

Chapter 6 outlined the kinds of ecoregions that occur throughout the world. This chapter examines in greater detail the ecoregions of the United States at the domain and division levels. By applying the same classification and criteria, we obtain the picture shown in Figure 7.1. Subdivisions, called provinces, are delineated and described elsewhere (Bailey 1995).

## 100 Polar Domain

Climates of the polar domain, located at high latitudes, are controlled chiefly by polar and arctic air masses. In general, climates in the polar domain have low temperatures, severe winters, and small amounts of precipitation, most of which falls in the summer. Polar systems are dominated by a periodic fluctuation of solar energy and temperature, in which the annual range is far greater than the diurnal range. The intensity of the solar radiation is never very high compared with ecosystems of the middle latitudes and tropics.

In areas where summers are short and temperatures are generally low throughout the year, temperature efficiency rather than effectiveness of precipitation is the critical factor in plant distribution and soil development. Two major divisions have been recognized and delimited in terms of temperature efficiency—the tundra and the subarctic (tayga). Climate diagrams in Figure 7.2 provide general information on the character of the climate in these two divisions.

R.G. Bailey, *Ecosystem Geography*, DOI 10.1007/978-0-387-89516-1_7, © Springer Science+Business Media, LLC 2009

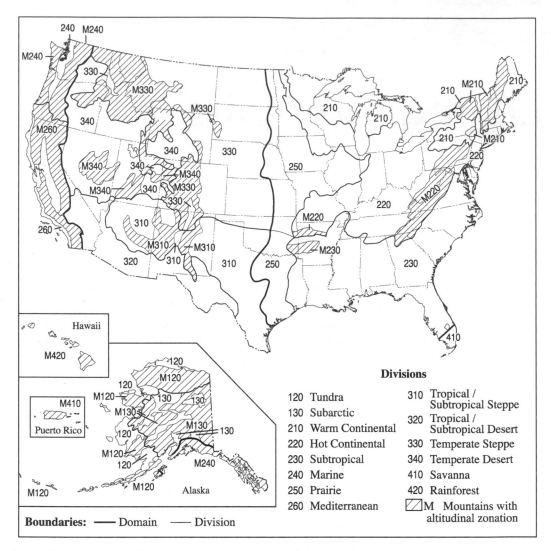

**Figure 7.1.** Second-order ecosystem regionalization of the United States. Boundaries are approximate. From Bailey (1995).

## 120 Tundra Division

The northern continental fringes of North America from the Arctic Circle northward to about the 75th parallel lie within the outer zone of control of arctic air masses. This produces the tundra climate that Trewartha (1968) designated by symbol *Ft*. Average temperature of the warmest month lies between 10°C and 0°C.

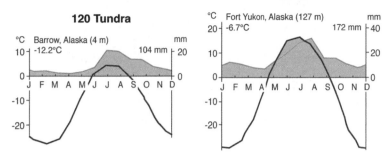

**Figure 7.2.** Climate diagrams from the tundra and from the extremely continental boreal (tayga) regions of Alaska. Redrawn from Walter et al. (1975).

The tundra climate has very short, cool summers and long, severe winters (see Fig. 7.2, climate diagram for Barrow, Alaska). No more than 188 days per year, and sometimes as few as 55, have a mean temperature higher than 0°C. Annual precipitation is light, often less than 200 mm, but because potential evaporation is also very low, the climate is humid.

Vegetation on the tundra consists of grasses, sedges, and lichens, with willow shrubs (Fig. 7.3). Farther south, the vegetation changes into birch-lichen woodland, then into a needleleaf forest. In some places, a distinct tree line separates the forest from tundra. Köppen (1931) used this line, which coincides approximately with the 10°C isotherm of the warmest month, as a boundary between subarctic and tundra climates.

Soil particles of tundra derive almost entirely from mechanical breakup of the parent rock, with little or no chemical alteration. Inceptisols, soils with weakly differentiated horizons, dominate. Continual freezing and thawing of the soil has disintegrated its particles. As in the northern continental interior, the tundra has a permanently frozen sublayer of soil known as permafrost. The permafrost layer is more than 300 m thick throughout the region; seasonal thaw reaches only 10–60 cm below the surface.

Geomorphic processes are distinctive in the tundra, resulting in a variety of curious landforms. Under a protective layer of sod, water in the soil melts in summer to produce a thick mud that sometimes flows downslope to create bulges, terraces, and lobes on hillsides. The freeze and thaw of water in the soil also sorts the coarse particles from the fine particles, giving rise to such patterns in the ground as rings, polygons, and stripes made of stone. The coastal plains have many lakes of thermokarst origin, formed by melting groundwater.

**Figure 7.3.** Tundra in Alaska. Photograph by USDA Forest Service.

## 130 Subarctic Division

The source region for the continental polar air masses is south of the tundra zone between latitudes 50° and 70° N. The climate type here shows great seasonal range in temperature. Winters are severe, and the region's small amounts of annual precipitation are concentrated in the 3 warm months. This cold, snowy, forest climate, referred to in this volume as the boreal subarctic type, is classified as $E$ in the Köppen-Trewartha system. This climate is moist all year, with cool, short summers (see Fig. 7.2, climate diagram for Fort Yukon, Alaska). Only 1 month of the year has an average temperature above 10°C.

Winter is the dominant season of the boreal subarctic climate. Because average monthly temperatures are subfreezing for 6–7 consecutive months, all moisture in the soil and subsoil freezes solidly to depths of

**Figure 7.4.** Patterned ground in a muskeg area northeast of Fort Yukon, Alaska. Photograph by T.G. Freeman, Soil Conservation Service.

a few meters. Summer warmth is insufficient to thaw more than a meter or so at the surface, so permafrost prevails under large areas (Fig. 7.4). Seasonal thaw penetrates from 0.5 to 4 m, depending on latitude, aspect, and kind of ground. Despite the low temperatures and long winters, the valleys of interior Alaska were not glaciated during the Pleistocene, probably because of insufficient precipitation.

The subarctic climate zone coincides with a great belt of needleleaf forest, often referred to as boreal forest, and open lichen woodland known as tayga. Most trees are small, with less value for lumber than for pulpwood.

The arctic needleleaf forest grows on Inceptisols with pockets of wet organic Histosols. These light gray soils are wet, strongly leached, and acidic. A distinct layer of humus and forest litter lies beneath the top soil layer. Agriculture potential is poor, due to natural infertility of soils and the prevalence of swamps and lakes left by the departed ice sheets. In some places, ice scoured the rock surfaces bare, entirely stripping off the overburden. Elsewhere rock basins were formed and stream courses dammed, creating countless lakes.

# 200 Humid Temperate Domain

The climate of the humid temperate domain, located in the midlatitudes (30°–60°), is governed by both tropical and polar air masses. The midlatitudes are subject to cyclones; much of the precipitation in this belt comes from rising moist air along fronts within those cyclones. Pronounced seasons are the rule, with strong annual cycles of temperature and precipitation. The seasonal fluctuation of solar energy and temperature is greater than the diurnal. The climates of the midlatitudes have a distinctive winter season, which tropical climates do not.

The humid temperate domain contains forests of broadleaf deciduous and needleleaf evergreen trees. The variable importance of winter frost determines six divisions: warm continental, hot continental, subtropical, marine, prairie, and Mediterranean. Climate diagrams for these divisions are presented in Figure 7.5.

## 210 Warm Continental Division

South of the eastern area of the subarctic climate, between latitudes 40° and 55° N and from the continental interior to the east coast, lies the humid, warm-summer, continental climate. Located squarely between the source regions of polar continental air masses to the north and maritime or continental tropical air masses to the south, it is subject to strong seasonal contrasts in temperature as air masses push back and forth across the continent.

The Köppen-Trewartha system designates this area as *Dcb*, a cold, snowy, winter climate with a warm summer (see Fig. 7.5, climate diagram for Iron Mountain, Michigan). This climate has 4–7 months when temperatures exceed 10°C, with no dry season. The average temperature during the coldest month is below 0°C. The warm summer signified by the letter *b* has an average temperature during its warmest month that never exceeds 22°C. Precipitation is ample all year but substantially greater during the summer.

Needleleaf and mixed needleleaf-deciduous forest grows throughout the colder northern parts of the humid continental climate zone, extending into the mountain region of the Adirondacks and northern New England. Here soils are Spodosols. Such soils have a low supply of bases and a horizon in which organic matter and iron and aluminum have accumulated. They are strongly leached but have an upper layer of humus. Cool temperatures inhibit bacterial activity that would destroy this organic matter in tropical regions. Soils are deficient in calcium, potassium, and magnesium and are generally acidic. Thus, they are poorly suited to crop production, even though adequate rainfall is generally ensured. Conifers thrive there.

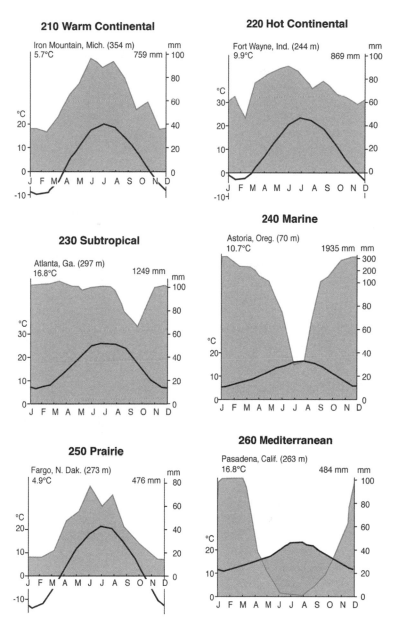

**Figure 7.5.** Climate diagrams from the deciduous forest regions (continental with hot and warm summers) and the mixed deciduous-coniferous forest region (more moderate), from the coniferous forest regions (winter rains and no summer drought), from the prairie region (cold winter but hot summer), and from the sclerophyllous regions of California (dry summer). Redrawn from Walter et al. (1975).

## 220 Hot Continental Division

South of the warm continental climate lies another division in the humid tropical domain, one with a humid, hot-summer continental climate. It has the same characteristics as the warm continental except that it is more moderate and has hot summers and cool winters (see Fig. 7.5, climate diagram for Fort Wayne, Indiana). The boundary between the two is the isotherm of 22°C for the warmest month. In the warmer sections of the humid temperate domain, the frost-free or growing season continues for 5–6 months, in the colder sections only 3–5 months. Snow cover is deeper and lasts longer in the northerly areas.

In the Köppen-Trewartha system, areas in this division are classified as *Dca* (*a* signifies hot-summer). We include in the hot continental division the northern part of Köppen's *Cf* (subtropical) climate region in the eastern United States. Köppen used as the boundary between the *C-D* climates, the isotherm of –3°C for the coldest month. Thus, for example, Köppen places New Haven, Connecticut, and Cleveland, Ohio, in the same climatic region as New Orleans, Louisiana, and Tampa, Florida, despite obvious contrasts in January mean temperatures, soil groups, and natural vegetation between these northern and southern zones. Trewartha (1968) redefined the boundary between *C* and *D* climates as the isotherm of 0°C of the coldest month, thereby pushing the climate boundary south to a line extending roughly from St. Louis to New York City. Trewartha's boundary is adopted here to distinguish between humid continental and humid subtropical climates.

Natural vegetation in this climate is winter deciduous forest, dominated by tall broadleaf trees that provide a continuous dense canopy in summer but shed their leaves completely in the winter (Fig. 7.6). Lower layers of small trees and shrubs are weakly developed. In spring, a luxuriant ground cover of herbs quickly develops but is greatly reduced after trees reach full foliage and shade the ground.

Soils are chiefly Inceptisols, Ultisols, and Alfisols, rich in humus and moderately leached with a distinct, light-colored, leached zone under the upper dark layer. The Ultisols have a low supply of bases and a horizon of accumulated clay. Where topography is favorable, diversified farming and dairying are the most successful agricultural practice.

## 230 Subtropical Division

The humid subtropical climate, marked by high humidity (especially in summer) and the absence of really cold winters, prevails throughout the Southern Atlantic and Gulf Coast States.

**Figure 7.6.** An oak-hickory forest in Ohio. Photograph by Robert K. Winters, U.S. Forest Service.

In the Köppen-Trewartha system, this area lies within the *Cf* climate, described as temperate and rainy with hot summers (see Fig. 7.5, climate diagram for Atlanta, Georgia). The *Cf* has no dry season; even the driest summer month receives at least 30 mm of rain. The average temperature of the warmest month is warmer than 22°C. Rainfall is ample all year but is markedly greater during summer. Thunderstorms, whether of thermal, squall-line, or cold-front origin, are especially frequent in summer. Tropical cyclones and hurricanes strike the coastal area occasionally, always bringing heavy rains. Winter fronts bring precipitation, some in the form of snow. Temperatures are moderately wide in range, comparable with those in tropical deserts, but without the extreme heat of a desert summer.

Soils of the moister, warmer parts of the humid subtropical regions are strongly leached Ultisols related to those of the humid tropical and equatorial climates. Rich in oxides of both iron and aluminum, these soils are poor in many of the plant nutrients essential for successful agricultural production.

Forest is the natural vegetation throughout most areas of this division. Much of the sandy coastal region of the southeastern United States today is covered by a second growth forest of longleaf (*Pinus palustris*), loblolly

(*Pinus taeda*), and slash pines (*Pinus elliotii*). Inland areas have deciduous forest.

## 240 Marine Division

Situated on the Pacific Coast between latitudes 40° and 60° N is a zone
that receives abundant rainfall from maritime polar air masses and has a
narrow range of temperature because it borders on the ocean.

Trewartha (1968) classified the marine west coast climate as *Do*—
temperate and rainy, with warm summers. The average temperature of
the warmest month is below 22°C, but at least 4 months of the year
have an average temperature of 10°C. The average temperature during
the coldest month of the year is above 0°C. Precipitation is abundant
throughout the year but is markedly reduced during the summer (see
Fig. 7.5, climate diagram for Astoria, Oregon). Although total rainfall is
not great by tropical standards, the cooler air temperatures reduce evaporation and produce a damp, humid climate with much cloud cover.
Mild winters and relatively cool summers are typical. Coastal mountain
ranges influence precipitation markedly in these middle latitudes. The
mountainous coasts of British Columbia and Alaska annually receive
1530–2040 mm of precipitation and more. Heavy precipitation greatly
contributed to the development of fiords along the coast. Heavy snows in
the glacial period fed vigorous valley glaciers that descended to the seas,
scouring deep troughs that reach below sea level at their lower ends.

Needleleaf forest is the natural vegetation of the marine division. In the
coastal ranges of the Pacific Northwest, Douglas-fir, western red-cedar
(*Thuja plicata*), and spruce grow to enormous heights, forming some
of the densest of all coniferous forest with some of the world's largest
trees.

Soils are strongly leached, acidic Inceptisols and Ultisols. Due to the
region's cool temperatures, bacterial activity is slower than in the warm
tropics, so unconsumed vegetative matter forms a heavy surface deposit.
Organic acids from decomposing vegetation react with soil compounds,
removing bases such as calcium, sodium, and potassium.

## 250 Prairie Division

Prairies are typically associated with continental, midlatitude climates
designated as *subhumid*. Precipitation in these climates ranges from 510
to 1020 mm per year and is almost entirely offset by evapotranspiration (see Fig. 7.5, climate diagram for Fargo, North Dakota). In summer, air and soil temperatures are high. Soil moisture in the uplands is

inadequate for tree growth, and deeper sources of water are beyond the reach of tree roots. Prairies form a broad belt extending from Texas northward to southern Alberta and Saskatchewan. Forest and prairie mix in a transitional belt on the eastern border of the division.

The prairie climate is not designated as a separate variety in the Köppen-Trewartha system. Geographers' recognition of the prairie climate (Thornthwaite 1931; Borchert 1950) has been incorporated into the system presented here. Prairies lie on the arid western side of the humid continental climate, extending into the subtropical climate at lower latitudes. Temperature characteristics correspond to those of the adjacent humid climates, forming the basis for two types of prairies: temperate and subtropical.

Tallgrasses associated with subdominant broad-leaved herbs dominate prairie vegetation. Trees and shrubs are almost totally absent, but a few may grow as woodland patches in valleys and other depressions. Deeply rooted grasses form a continuous cover. They flower in spring and early summer, the forbs in late summer. In the tallgrass prairie of Iowa, for example, typical grasses are big bluestem (*Andropogon gerardii*) and little bluestem (*Schizachyrium scoparium*); a typical forb is black-eyed Susan (*Rudbeckia hirta*).

Because rain falls less in the grasslands than in forest, less leaching of the soil occurs. The pedogenic process associated with prairie vegetation is calcification, as carbonates accumulate in the lower layers. Soils of the prairies are Mollisols, which have black, friable, organic surface horizons and a high content of bases. Grass roots deeply penetrate these soils. Bases brought to the surface by plant growth are released on the surface and restored to the soil, perpetuating fertility. These soils are the most productive of the great soil groups.

## 260 Mediterranean Division

Situated on the Pacific Coast between latitudes 30° and 45° N is a zone subject to alternately wet and dry seasons, the transition zone between the dry west coast desert and the wet west coast.

Trewartha (1968) classified the climate of these lands as *Cs*, signifying a temperate rainy climate with dry, hot summers. The symbol *s* signifies a dry summer (see Fig. 7.5, climate diagram for Pasadena, California). The combination of wet winters with dry summers is unique among climate types and produces a distinctive natural vegetation of hard-leaved evergreen trees and shrubs called sclerophyll forest. Various forms of sclerophyll woodland and scrub are also typical (Fig. 7.7). Trees and shrubs must withstand the severe summer drought—2–4 rainless months—and severe evaporation.

**Figure 7.7.** Sclerophyll open woodland south of San Francisco. Most of the trees are oaks. Photograph by R.E. Wallace, U.S. Geological Survey.

Soils of this Mediterranean climate are not susceptible to simple classification. Alfisols and Mollisols typical of semiarid climates are generally found.

# 300 Dry Domain

The essential feature of a dry climate is that annual losses of water through evaporation at the earth's surface exceed annual water gains from precipitation. Due to the resulting water deficiency, no permanent streams originate in dry climate zones. Because evaporation, which depends chiefly on temperature, varies greatly from one part of the earth to another, no specific value for precipitation can be used as the boundary for all dry climates. For example, 610 mm of annual precipitation may produce a humid climate and forest cover in cool northwestern Europe, but the same amount in the hot tropics produces semiarid conditions.

We commonly recognize two divisions of dry climates: the arid desert and the semiarid steppe. Generally, the steppe is a transitional belt surrounding the desert and separating it from the humid climates beyond. The boundary between arid and semiarid climates is arbitrary but is commonly defined as one-half the amount of precipitation separating steppe from humid climates. These climates are displayed in Figure 7.8.

Of all the climatic groups, dry climates are the most extensive; they occupy one-fourth or more of the earth's land surface (see Fig. 5.13).

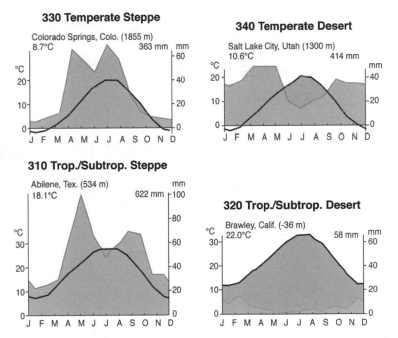

**Figure 7.8.** Climate diagrams of steppe and desert stations: (*above*) with summer rain, and some rain at all seasons; (*below*) with summer rain (with dry season) and from the semidesert sagebrush region with long summer drought. Redrawn from Walter et al. (1975).

## 310 Tropical/Subtropical Steppe Division

Tropical steppes border the tropical deserts on both the north and south and in places on the east as well. Locally, altitude causes a semiarid steppe climate on plateaus and high plains that would otherwise be desert. Steppes on the poleward fringes of the tropical deserts grade into the Mediterranean climate in many places. In the United States, they are cut off from the Mediterranean climate by coastal mountains, which allow the tropical deserts to extend farther north.

Trewartha (1968) classified the climate of tropical/subtropical steppes as *BSh*, indicating a hot semiarid climate where potential evaporation exceeds precipitation and where all months have temperatures above 0°C (see Fig. 7.8, climate diagram for Abilene, Texas).

Steppes typically are grassland of shortgrasses and other herbs and with locally developed shrub and woodland. Pinyon-juniper woodland (*Pinus-Juniperus*) grows on the Colorado Plateau, for example. To the east, in New Mexico and Texas, the grasslands grade into savanna woodland or semideserts (Fig. 7.9) composed of xerophytic shrub and

**Figure 7.9.** Subtropical semidesert in New Mexico. Sand sage (*Artemisia filifolia*) is the principal shrub in the picture. Photograph by John McConnell, Soil Conservation Service.

trees, and the climate becomes nearly arid-subtropical. Cactus plants are present in some places. These areas support limited grazing but are not generally moist enough for crop cultivation without irrigation. Soils are commonly Mollisols and Aridisols, containing some humus.

## 320 Tropical/Subtropical Desert Division

The continental desert climates are south of the Arizona-New Mexico mountains. They are not only extremely arid but also have extremely high air and soil temperatures. Direct solar radiation is very high, as is outgoing radiation at night, causing extreme variations between day and night temperatures and a rare nocturnal frost. Annual precipitation is less than 200 mm and less than 100 mm in extreme deserts (see Fig. 7.8, climate diagram for Brawley, California). These areas have climates that Trewartha (1968) calls *BWh*.

**Figure 7.10.** Desert vegetation in Death Valley. Photograph by National Park Service.

Dry-desert vegetation characterizes the region. Widely dispersed xerophytic plants provide negligible ground cover. In dry periods, visible vegetation is limited to small hard-leaved or spiny shrubs, cacti, or hard grasses. Many species of small annuals may be present, but they appear only after the rare but heavy rains have saturated the soil.

In the Mojave-Sonoran Deserts (American Desert), plants are often so large that some places have a near-woodland appearance. They include the treelike saguaro cactus (*Cereus giganteus*), the prickly pear cactus (*Opuntia*), the ocotillo (*Fouquieria splendens*), creosote bush (*Larrea tridentata*), and smoke tree (*Dalea spinosa*). However, much of the desert of the southwestern United States is, in fact, scrub, thorn scrub, savanna, or steppe grassland. Parts of this region have no visible plants. They are made up of shifting dune sand or almost sterile salt flats (Fig. 7.10).

The dominant pedogenic process is salinization, which produces areas of salt crust where only salt-loving plants (halophytes) can survive. Calcification is conspicuous on well-drained uplands, where encrustations and deposits of calcium carbonate (caliche) are common. Humus is lacking, and soils are mostly Aridisols and dry Entisols.

## 330 Temperate Steppe Division

Temperate steppes are areas that have a semiarid continental climatic regime in which, despite maximum summer rainfall, evaporation usually exceeds precipitation. Trewartha (1968) classified the climate as *BSk*. The letter *k* signifies a cool climate with at least 1 month of average temperature below 0°C. Winters are cold and dry, summers warm to hot (see Fig. 7.8, climate diagram for Colorado Springs, Colorado). Drought periods are common in this climate. With the droughts come the dust storms that blow the fertile topsoil from vast areas of plowed land being used for dry farming (Fig. 7.11).

The vegetation is steppe, sometimes called shortgrass prairie, and semidesert. Typical steppe vegetation consists of many species of shortgrasses that usually grow in sparsely distributed bunches. Many species of grasses and other herbs occur. Buffalograss (*Buchlow dactyloides*) is typical grass of the American steppe. Other typical plants are the sunflower (*Helianthus annuus*) and locoweed (*Oxytropis*). Scattered shrubs and low trees sometimes grow in the steppe; all gradations of cover are present, from semidesert to woodland. Because ground cover is generally sparse, much soil is exposed.

The semidesert cover is xerophytic shrub vegetation accompanied by a poorly developed herbaceous layer. Trees are generally absent. An exam-

**Figure 7.11.** A dust storm approaching in the steppe of eastern Colorado. Photograph by Soil Conservation Service.

ple of semidesert cover is the sagebrush vegetation of the middle and southern Rocky Mountain region and the Colorado Plateau.

In this climatic regime, the dominant pedogenic process is calcification, with salinization on poorly drained sites. Soils contain an excess of precipitated calcium carbonate and are rich in bases. Mollisols are typical in steppe lands. The soils of the semidesert shrub are Aridisols, with little organic content, and (occasionally) clay horizons, and (in some places) accumulations of various salts. Humus content is small because the vegetation is so sparse.

## 340 Temperate Desert Division

Temperate deserts of continental regions have low rainfall and strong temperature contrasts between summer and winter. In the intermountain region of the western United States, between the Pacific and Rocky Mountains, the temperate desert has characteristics of a sagebrush semidesert, with a pronounced drought season and a short humid season.

Most precipitation falls in the winter, despite a peak in May (see Fig. 7.8, climate diagram of Salt Lake City, Utah). Aridity increases markedly in the rain shadow of the Pacific mountain ranges. Even at intermediate elevations, winters are long and cold, with temperatures below 0°C.

Under the Köppen-Trewartha system, this is the true desert, *BWk*. The letter *k* signifies that at least 1 month has an average temperature below 0°C. These deserts differ from those at lower latitude chiefly in their far greater annual temperature range and much lower winter temperatures. Unlike the dry climates of the tropics, middle-latitude dry climates receive a portion of their precipitation as snow.

Temperate deserts support the sparse xerophytic shrub vegetation typical of semideserts. One example is the sagebrush vegetation of the Great Basin and northern Colorado Plateau region. Recently, semidesert shrub vegetation seems to have invaded wide areas of the western United States that were formerly steppe grasslands, due to overgrazing and trampling by livestock. Soils of the temperate desert are Aridisols, low in humus and high in calcium carbonate. Poorly drained areas develop saline soils, and salt deposits cover dry lake beds.

# 400 Humid Tropical Domain

Equatorial and tropical air masses largely control the humid tropical group of climates found at low latitudes. Every month of the year has an average temperature above 18°C, and no winter season occurs. In these tropical systems, the primary periodic energy flux is diurnal: the temperature variation from day to night is greater than from season to season. Average annual rainfall is heavy and exceeds annual evaporation but varies in amount and in season and distribution.

Two types of climates are differentiated on the basis of the seasonal distribution of precipitation (Fig. 7.12). Tropical wet (or rainforest) climate has ample rainfall through 10 or more months of the year. Tropical wet-dry, or savanna, climate has a dry season more than 2 months long.

## 410 Savanna Division

The latitude belt between 10° and 30° N is intermediate between the equatorial and middle-latitude climates. This produces the tropical wet-dry savanna climate, which has a wet season controlled by moist, warm, maritime, tropical air masses at times of high sun and a dry season controlled by the continental tropical masses at times of low sun (see

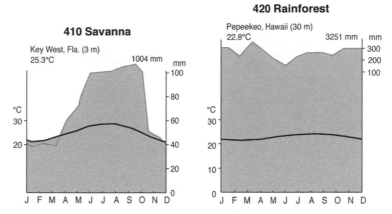

**Figure 7.12.** Climate diagrams of savanna and rainforest stations: (*left*) with maximum rain during the high-sun period; (*right*) with constantly wet climate. Redrawn from Walter et al. (1975).

Fig. 7.12, diagram for Key West, Florida). Trewartha (1968) classified the tropical wet-dry climate as *Aw*, the letter *w* signifying a dry winter.

Alternating wet and dry seasons result in the growth of a distinctive vegetation known generally as tropical savanna. It is characterized by open expanses of tallgrasses, interspersed with hardy, drought-resistant shrubs and trees. Some areas have savanna woodland, monsoon forest, thornbush, and tropical scrub. In the dry season, grasses wither into straw, and many tree species shed their leaves. Other trees and shrubs have thorns and small or hard, leathery leaves that resist loss of water.

Soils are mostly Histosols and Inceptisols. Heavy rainfall and high temperatures cause heavy leaching. Streamflow in these regions is subject to strong seasonal fluctuations, in striking contrast to the constant streamflow typical of rainforest climates. In the rainy season, extensive low-lying areas are submerged; in the dry season, streamflow dissipates, exposing channel bottoms of sand and gravel as stream channels and mud flats dry out.

In North America, the savanna division is found in southern Florida (Fig. 7.13), where fluctuating water levels strongly influence habitats and fauna. Large numbers of birds are characteristic.

## 420 Rainforest Division

The wet equatorial or rainforest climate lies between the equator and latitude 10° N. Average annual temperatures are close to 27°C; seasonal variation is virtually imperceptible. Rainfall is heavy throughout the year,

**Figure 7.13.** Flat, marshy surface of the Florida Everglades. Photograph by Jack Boucher, National Park Service.

but the monthly averages differ considerably due to seasonal shifts in the equatorial convergence zone, and a consequent variation in air-mass characteristics (see Fig. 7.12, climate diagram for Pepeekeo, Hawaii). Trewartha (1968) defines this climate as *Ar*, with no month averaging less than 60 mm of rainfall.

The equatorial region has a rainforest, or selva type of vegetation, unsurpassed in number of species and luxuriant tree growth. Broadleaf trees rise 30–45 m in height, forming a dense leaf canopy through which little sunlight can reach the ground. Giant lianas (woody vines) hang from trees. The forest is mostly evergreen, but individual species have various leaf-shedding cycles.

Rainforests are home to small forest animals able to live and travel in the continuous forest canopy. Bird species are numerous and spectacularly plumaged.

Copious rainfall and high temperatures combine to keep chemical processes continuous on the rocks and soils. Leaching of all soluble elements of the deeply decayed rock produces Ultisols and Oxisols that are often especially rich in hydroxides of iron, magnesium, and aluminum.

Streamflow is fairly constant because the large annual water surplus provides ample runoff. Dense vegetation lines river channels. Sand bars and sand banks are less conspicuous than in drier regions. Floodplains have cutoff meanders (oxbows) and many swampy sloughs where meandering river channels have shifted their courses. Although water is abundant, river systems carry relatively little dissolved material because thorough leaching of soils removes most soluble mineral matter before it reaches steams.

Not all equatorial rainforest areas have low relief. Hilly or mountainous belts have very steep slopes; frequent earthflows, slides, and avalanches of soil and rock strip surfaces down to bedrock.

# Mountains with Altitudinal Zonation

Mountains are distinguished where, as a result of the influence of altitude, the climatic regime differs substantially from adjacent lowlands to cause vertical climate-vegetation-soil zonation.

The succession of altitudinal belts in the Front Range of the Rocky Mountains near Colorado Springs provides an example of a mountain ecoregion in a temperate-steppe climatic regime (M330; Fig. 7.14). The shortgrass prairie at the foot of the mountains is succeeded at 1500 m by a belt with tallgrass prairie, and then by a belt only 50 m wide of deciduous shrubs, with *Pinus edulis* and *Juniperus* (pinyon belt). Next come the forest belts in which *Pinus ponderosa, Pseudotsuga menziesii,* and *Picea engelmannii,* successively, achieve dominance. The timberline is reached at 3700 m, and above a narrow subalpine belt with dwarf *Picea* and *Dasiphora (Pontintilla) fruticosa* bushes, the alpine belt is reached.

# American Ecoregions in Review

In this chapter, we have surveyed the ecoregion geography of the country at two levels—domain and division. These levels are based on macrofeatures of the climate. Each type of climate, together with its characteristic natural vegetation, soils, landforms, and geomorphic processes, comprises a unique ecoregion and supports a distinctive pattern of ecosystems.

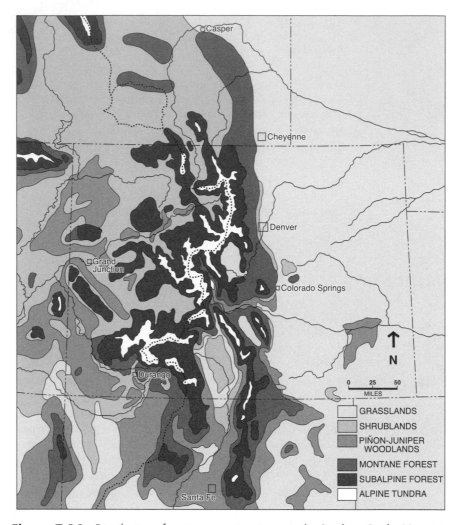

**Figure 7.14.** Distribution of major ecosystem types in the Southern Rocky Mountain region. From Mutel and Emerick (1992); reproduced with permission.

The above vegetation descriptions are those of *potential* natural vegetation, which may not exist under current land-use practices. Furthermore, the descriptions refer to zonal conditions (i.e., the vegetation that would occur if the area were flat and well drained). Local contrast (related to elevation, geology, or groundwater) can exist within a region forming a complex intraregional mosaic. We discuss these contrasts in the following chapters.

# CHAPTER 8

# Ecoregion Redistribution Under Climate Change

Ecoregions are large, regional scale ecosystems—such as the Sonoran Desert. These regions are primarily defined by climatic conditions and on the prevailing plant formations determined by those conditions. Climate, as a source of energy and water, acts as the primary control for ecosystem distribution, including ecoregions. As climate changes, so do ecosystems, as a petrified forest lying in a desert attests (Fig. 8.1). With recognition that climate is a principal controlling factor for ecosystems, there exists a need to study potential climatic change in terms of its ramifications for the Earth's terrestrial ecosystems. Knowing where ecological shifts will most likely occur and consequences associated with such shifts are prerequisite to the evaluation of these changes in terms of development and resource management decisions.

## Long-Term Climate Change

The distribution of plant and animal communities, and indeed of entire ecoregions, has varied tremendously with past changes in climate, even in the absence of man's activities. The spatial distribution of life forms today as a function of latitude, continental position, and elevation looks very different compared to that of 5000 or 10,000 years before the present (BP).

Climatic changes on Earth during the past 500,000 years have been dramatic (Fig. 8.2). Each glacial–interglacial cycle is about 100,000 years in duration, with 90,000 years of gradual climatic cooling followed by rapid warming and 10,000 years of interglacial warmth. The peak of the last glacial period, or ice age, was about 18,000 years BP and ended approximately 10,000 years BP.

R.G. Bailey, *Ecosystem Geography*, DOI 10.1007/978-0-387-89516-1_8,
© Springer Science+Business Media, LLC 2009

**Figure 8.1.** Petrified forest in a desert zone in Arizona. Photograph: D.B. Sterrett, U.S. Geological Survey.

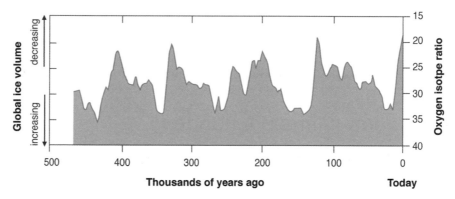

**Figure 8.2.** Changes in glacial history derived from the evidence of deep sea cores obtained from the Indian Ocean. From Imbrie and Imbrie (1979), p. 169; reproduced with permission.

During the glacial periods, the ice caps of the world were greatly expanded. On the periphery of the great ice sheets, there were great areas of open tundra frequently underlain by permafrost. The areas of forest that form the natural vegetation of much of north and eastern North America as well as western Europe today were largely occupied by cold, rather dry tundra and steppe. A representation of the expansion and contraction of ecoclimatic zones is given in Figure 8.3, which shows the migration of zonal belts in relation to glacial advance and retreat. There is evidence that, at certain times in the past, there has been more water in low-latitude desert areas: huge lakes, for example, filled the now largely dry basins of the southwest United States. However, there is also evidence that, in other areas, the glacial periods were characterized not by increased humidity but by reduced precipitation. The most spectacular evidence for this is the great expansion of sand dunes in low latitudes: studies of air photos and satellite imagery indicate that degraded ancient dunes lie in areas that are now quite moist. Today, about 10% of the land area between 30° N and 30° S is covered by active sand deserts. At the time of the last great glacial advance, about 18,000 years ago, such deserts probably characterized almost 50% of the land area at those latitudes. In this period, tropical rainforests and adjacent savannas were reduced to a narrow corridor and were much less extensive than they are today.

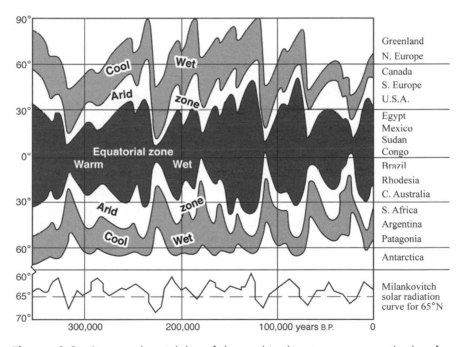

**Figure 8.3.** Suggested variability of the earth's climatic zones over the last few hundred thousand years. From Fairbridge 1963.

As to the causes of this variability: it is evident that the changing configuration and position of the continents and oceans (caused by plate tectonics), together with the uplift of mountain chains, have had a major effect on world climate. For most of the Earth's history, the continents and oceans have been so arranged that warm ocean currents from the tropics were able to flow easily into the northern and southern polar regions; there were no barriers to the flow of currents from low to high latitudes. This is not the situation today. At present, a continent (Antarctica) covers a large area centered on the South Pole; the Arctic Ocean, centered on the North Pole, is almost cut off from surrounding oceans because of the arrangement of northern continents around it. Because land barriers prevent warm ocean currents from circulating the sun's energy away from tropical and temperate latitudes toward polar ones, we are now experiencing a glacial age.

Another geological change that can affect world climates is volcanic activity. A prolonged period of severe eruptions can inject large amounts of dust into the atmosphere. This might decrease the quantity of solar radiation reaching the Earth's surface and so cause a phase of cooling. A meteorite that struck the Earth at the end of the Cretaceous period (65 M years BP) is thought to have had that effect and initiated the extinction of the dinosaurs.

It is also possible that the output of solar radiation from the sun varies through time, possibly in a cyclic manner. There is some good evidence of sunspot activity having a cyclic pattern, with 11- and 22-year cycles being one particularly noted.

What has become more certain in recent years, however, is that the amount of radiation received at the Earth's surface through time has varied as a consequence of the ever changing position of the Earth with respect to the sun. This is called the Milankovitch cycle after its discoverer. The basic idea is that there are three ways in which the Earth's position varies (Fig. 8.4). First, the Earth's orbit around the sun is not a perfect circle but an ellipse (a). This orbital eccentricity results in approximately 3.5% variation in the total amount of solar radiation received. Second, the tilt of the Earth's axis of rotation varies (b). And third, there is a mechanism which is based on the fact that the time of the year at which the Earth is nearest the sun varies (c). At times, the northern hemisphere is closest to the sun in winter; other times it is closest to the sun in summer. The reason for this is that the Earth wobbles like a slowing top and swivels its axis around.

The climatic effect of these cycles is a variation in the degree of contrast between summer and winter temperatures. When the contrast between seasons is comparatively slight, summer temperatures are not high enough for the previous winter's snow and ice to melt. Snow and ice accumulate, building up huge continental ice sheets in temperate latitudes. During another phase of the cycle when there are high summer temperatures, ice melts before the onset of the succeeding winter: the result is the advent of an interglacial period, such as the one now approaching (Fig. 8.5).

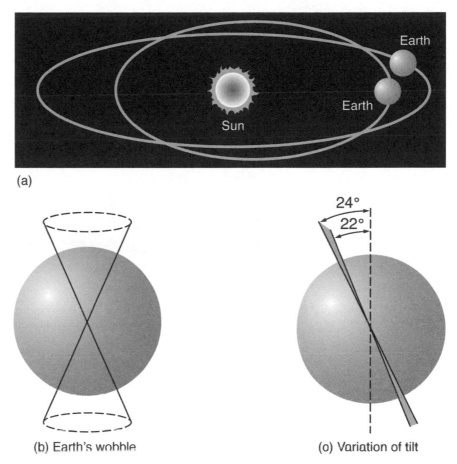

(a)

(b) Earth's wobble          (o) Variation of tilt

**Figure 8.4.** The three types of fluctuation in earth–sun relationship involved in the Milankovitch cycle. From Christopherson (2000), p. 517. From Christopherson, Robert W., *Geosystems: an introduction to physical geography*, 4E, © 2000, Pg. 517. Reprinted by permission of Pearson Education, Inc., Upper Saddle River, NJ.

# Use of the Köppen Climate Classification to Detect Climate Change

The Intergovernmental Panel on Climate Change science report (IPCC 2001) represents the consensus view on greenhouse-induced climatic changes expressed by the overwhelming majority of atmospheric scientists throughout the world. The reported equilibrium changes for

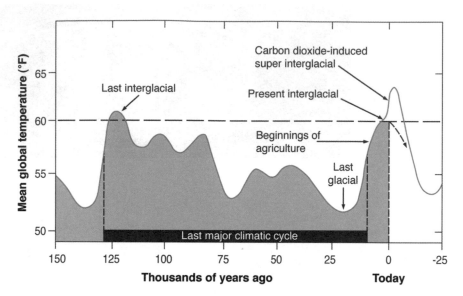

**Figure 8.5.** Global climate change over the past 150,000 years and projected for the next 25,000 years. A cooling trend is projected in the future based on the Milankovitch cycles, but this may be delayed by a warming period induced by elevated concentrations of carbon dioxide and other greenhouse gases in the atmosphere. From Mitchell (1977), p. 8.

doubling of $CO_2$ include temperature increases between 1.5°C and 4.5°C and global precipitation increases between +3 and +15%.

Boundaries of ecoregions coincide with certain climatic parameters. Based on macroclimatic conditions and on the prevailing plant formations determined by those conditions, I subdivided the continents into ecoregions with three levels of detail. Of these, the broadest, domains, and within them divisions, are based largely on the broad ecological zones of the German geographer Wladimir Köppen (1931; as modified by Trewartha 1968, Chapter 6). Zone boundaries take into account the near-surface air temperature and precipitation as the major variables with respect to their annual cycles and their linkages with natural vegetation patterns. Assignment is based on quantitative definitions and, as such, can be applied to any part of the Earth where climatic data are available. It is thereby possible to develop world maps for future climate simulated, for instance, under elevated atmospheric $CO_2$ concentrations.

The Köppen-Trewartha classification identified six main groups of climate, and all but one—the dry group—are thermally defined (see Table 6.1). They are as follows:

*Based on temperature criteria*

A. Tropical: Frost limits in continental locations; in marine areas 18°C for the coolest month

C. Subtropical: 8 months 10°C or above

D. Temperate: 4 months 10°C or above

E. Boreal: 1 (warmest) month 10°C or above

F. Polar: All months below 10°C

*Based on precipitation criteria*

B. Dry: Outer limits, where potential evaporation equals precipitation

The sensitivity of the Köppen climate classification to climatic change was tested by remapping the Köppen climate classes for an alternative climate by Kalvova et al. (2003). These investigators used the output of a series of general circulation models (GCM) to simulate climate for the period 1961–1990 and 2036–2065. Figure 8.6 summarizes the differences in the area distributions of the Köppen climate types. All GCM projections of warming climate (horizon 2050) show that the zones representing tropical rain climates (A) and dry climates (B) become larger, and the zones identified with boreal forest (D) and snow climates (E), together with the polar climates, are smaller. These results were similar to Lohmann et al. (1993), who did greenhouse gas warming simulations and found a retreat of regions of permafrost and the increase of areas with tropical rainy climates and dry climates.

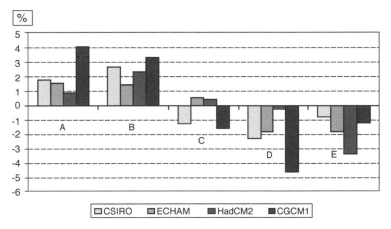

**Figure 8.6.** Differences in the coverage of different Köppen climate types between periods 1961–1990 and 2036–2065 for all GCMs. From Kalvova et al. (2003).

Ecological impacts of the recent warming trend in the arctic are already noted as changes in tree line and a decrease in tundra area with the replacement of ground cover by shrubs in northern Alaska and several locations in northern Eurasia. The poleward movement of Köppen's climate zones has been documented by Wang and Overland (2004). Figure 8.7 shows the spatial distributions of Köppen's climate classifications for two selected years. The left panels of Figure 8.7 are for 1978, the year with high tundra group coverage. By 1998, significant portions in the coverage of tundra group had been replaced by the boreal group. The coverage of tundra group being replaced by boreal group is further supported by Normalized Differences Vegetation Index (NDVI) data.

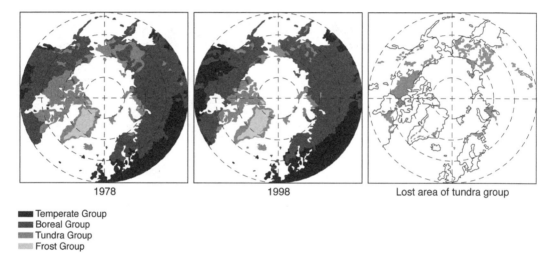

**Figure 8.7.** Spatial distribution of Köppen tundra climate classification for selected years. From Wang and Overland (2004).

Climate change may not only affect the boundaries between ecoregions. Figure 8.8 shows the predicted elevation shift of vegetation zones in the Great Basin in Nevada (temperate desert) that would occur assuming 3°C average climatic warming. The lower limit of woodland would shift approximately 500 m above its present elevation of 2280 m. This would decrease the area of woodland on all mountain ranges in the region and eliminate coniferous forest from some of them. Halpin (1994) cautions that changes in ecoclimatic zonation on elevational gradients cannot be explained by simple linear assumptions applied globally. There are significant latitudinal variation in the number of elevational zones present and their elevational limits (Chapter 5). For example, the

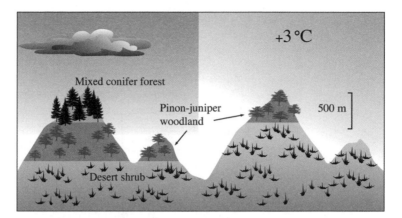

**Figure 8.8.** The approximate elevational boundaries of the vegetation types on the isolated mountain ranges of the Great Basin: (*left*) today; (*right*) in the future after a postulated climatic warming of approximately 3°C. From Brown (1995). From Brown. *Macroecology.* © 1995 The University of Chicago; reproduced with permission.

elevational limits of closed forest timberline, tree limit and krummholz zones vary significantly with latitudinal position of the mountain site. There is a distinct latitudinal trend with timberlines occurring at lower elevations with distance from the equator. Conceptual models of potential impacts of climate change must take into account differences in the elevational limits of zones at different latitudes.

Climate change can, in theory, cause a reduction in the spatial extent of a community. The alpine vegetation is one example of a community which will probably reduce in extent as direct result of climate change. Diaz and Eischeid (2007) analyzed changes in the Köppen "alpine tundra" climate classification type for the mountainous western United States by classifying 4-km pixels of topographically adjusted climate data in a geographic information system (GIS). There were 1226 4-km pixels classified as "alpine tundra" in the 1901–30 period, whereas in 1987–2006, there were only 336 thus categorized: a decline of ~73% (Fig. 8.9). Of particular note was that rising temperatures have caused the remaining classified alpine tundra in the last 20 years to be near the 10°C threshold for alpine tundra classification. Continuing warming past the threshold would imply that areas where this climate type is found today in the West will no longer be present.

This book uses a method to delineate ecosystems based on the Köppen climate classification system. It should be pointed out that different methods give different predictions about the future distribution of global vegetation (and therefore ecosystem) patterns (e.g., Emanuel et al. 1985; Prentice et al. 1992; Claussen and Esch 1994).

**Köppen Tundra 1901-1930**

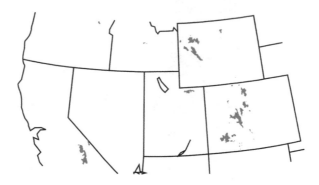

**Köppen Tundra 1987-2006**

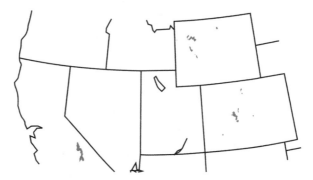

**Figure 8.9.** Distribution of Köppen classification "E" (tundra climates, E-T) corresponding to the "alpine tundra" climate in the western United States. From Diaz and Eischeid (2007).

# Summary

Climate acts as the primary control for ecosystem distribution, including ecoregions. As the climate varies, so do the ecoregions. Climatic changes on Earth during the past 500,000 years have been dramatic, resulting in ecoregion redistribution. The causes for long-term climatic change are attributed to changing configuration and position of the continents (caused by plate tectonics), together with uplift of mountain chains, volcanic activity, output of solar radiation, and amount of solar radiation received at the Earth's surface (Milankovitch cycles). Classifications that recognize the dependence of ecoregions on climate provide one means of constructing maps to display the impact of climate change on the geography of global ecoregions. A series of maps of the Köppen climate classification, as modified by Trewartha, was compared to Köppen map simulations based on models of climate under elevated atmospheric $CO_2$ concentrations. Results of studies show that the zones representing tropical rain climates and dry climates become larger and the zones identified with boreal forest and snow climates together with the polar climates becoming smaller. Climate change can cause a reduction in the spatial extent of a community. Montane coniferous forest and alpine tundra are examples of communities which will probably be reduced in extent as a direct result of climate change.

# CHAPTER 9

# Mesoscale: Landform Differentiation (Landscape Mosaics)

Macroclimate accounts for the largest share of systematic environmental variation at the macroscale or regional level. Within the same macroclimate, broad-scale landforms break up the east–west climatic pattern that would occur otherwise and provide a basis for further differentiation of ecosystems—the landscape mosaics mentioned earlier. The character of a landscape mosaic with identical geology will vary by the climatic zone. For example, vertical limestone would form quite different landscapes in a subarctic climate than in a hot and arid climate. Limestone in a subarctic climate occurs in depressions and shows intense karstification, whereas in hot and arid climates, it occurs in marked relief with a few cave tunnels and canyons inherited from colder Pleistocene time (Fig. 9.1).

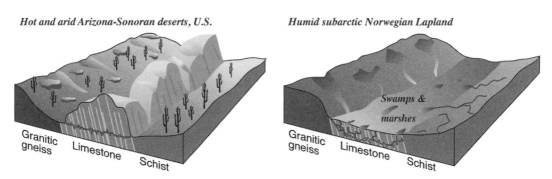

**Figure 9.1.** Landscape types resulting from similar geology in two different climatic regions. Redrawn from Corbel (1964).

R.G. Bailey, *Ecosystem Geography*, DOI 10.1007/978-0-387-89516-1_9,
© Springer Science+Business Media, LLC 2009

Landforms (with their geologic substrate, surface shape, and relief) influence place-to-place variations in ecological factors, such as water availability and exposure to radiant solar energy. Through varying height and degree of ground-surface inclination, landforms interact with climate and directly influence hydrologic and soil-forming processes.

Landform is the best correlation of vegetation and soil patterns at meso- and microscales. This is because landform controls the intensity of key factors important to plants and to the soils that develop with them (Hack and Goodlet 1960; Swanson et al. 1988). The importance of landform is apparent in several approaches to classification of forest land (e.g., Barnes et al. 1982). Even in areas of relatively little topographic relief, such as the glacial landforms of the upper Midwest of the United States, landform explains a great deal of the variability of ecosystems across the landscape (Host et al. 1987).

# Causes of Landscape Mosaic Pattern

Landscape mosaic patterns result, in part, from variability in landform. Geologic structure is a dominant control factor in the evolution of landforms at this scale. Take, for example, the Western Cordillera of North America (Malanson and Butler 2002): here, the geologic structure can be ascribed to plate tectonics where the North American and Pacific plates have collided giving rise to a series of block-faulted and folded mountain ranges (Fig. 9.2). The mountain ranges and intervening valleys are oriented in a north–south direction. The primary climatic gradients across mountain ranges are controlled by latitude and continental positions. However, the alignment of the ranges perpendicular to the primary zonal (east–west) flow of the atmosphere creates the other climatic gradient: a combination of orographic precipitation and rain shadow (see 5.24). These climatic gradients of energy and moisture are actually expressed at different scales. The energy gradient is generally a single latitudinal gradient for the entire cordillera, whereas the moisture gradient is spatially complex because of the repetition in the alternation of orographic uplift and rain shadow.

# Principal Landform Classes

Landforms come in all shapes and scales. On a continental scale within the same macroclimate, we commonly find several broad-scale landform patterns that break up the zonal patterns (Fig. 9.3). The landform

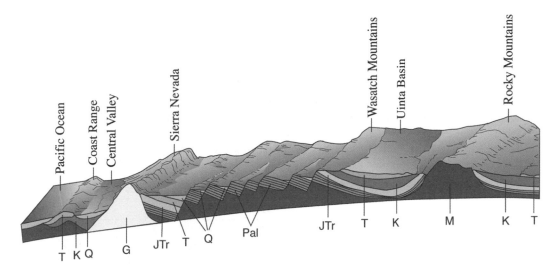

**Figure 9.2.** Geologic cross section of the western United States, showing control of the major topographic features: Q, Quaternary; T, Tertiary; K, Cretaceaous; JTr, Jurassic and Triassic; Pal, Paleozoic undivided; M, Metamorphic rocks, mostly Precambrian; G, granite. From Hunt (1967).

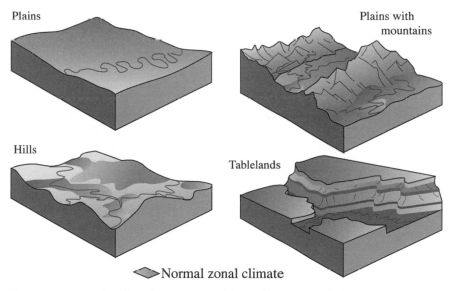

**Figure 9.3.** Land-surface form types and their effect on zonal climate.

classification of Edwin H. Hammond (1954, 1964), who classified land-surface forms in terms of existing surface geometry, is useful in determining the limits of various mesoecosystems or landscape mosaics. Landscape mosaics are made of multiple sites as described in Chapter 2. In the Hammond system, summarized in Table 9.1, landforms are identified on the basis of similarities and differences with respect to three major characteristics: relative amount of gently sloping land (less than 8%), local relief, and generalized profile (where and how much of the gently sloping land is located in the valley bottoms or in the uplands).

**Table 9.1.** Hammond's scheme of landform classification

| | | Symbol definition |
|---|---|---|
| Slope | | |
| | A | More than 80% of area gently sloping |
| | B | 50–80% of area gently sloping |
| | C | 20–50% of area gently sloping |
| | D | Less than 20% of area gently sloping |
| Local relief | | |
| | 1 | 0–30 m |
| | 2 | 30–90 m |
| | 3 | 90–150 m |
| | 4 | 150–300 m |
| | 5 | 300–900 m |
| | 6 | More than 900 m |
| Profile types | | |
| | a | More than 75% of gentle slope is in lowland |
| | b | 50–75% of gentle slope is in lowland |
| | c | 50–75% of gentle slope is on upland |
| | d | More than 75% of gentle slope is on upland |

On the basis of these characteristics alone, we may distinguish among (1) plains with a predominance of gently sloping land, coupled with low relief, (2) plains with some features of considerable relief, (3) hills with gently sloping land and low-to-moderate relief, and (4) mountains with little gently sloping land and high local relief.

On the basis of where the gently sloping land occurs in the profile, we may subdivide the second group into plains with hills, mountains, or tablelands. Approximate definitions of the grouping or generalized terrain types are as follows:

- nearly flat plains: A1; any profile

- rolling and irregular plains: A2, B1, B2; any profile

- plains with widely spaced hills or mountains: A3a or b, B3a or b to B6a or b
- partially dissected tablelands: B3c or d to B6c or d
- hills: D3, D4; any profile
- low mountains: D5; any profile
- high mountains: D6; any profile

Figure 9.4 gives examples of the principal terrain classes. Of course, much variety exists within these classes. Some plains, for instance, are flat and swampy, others rolling and well drained, and still others are simply broad expanses of smooth ice. Similarly, some mountains are low, smooth-sloped, and arranged in parallel ridges, whereas others are exceedingly high, with rugged, rocky slopes, glaciers, and snowfields.

To account for some of this variability, two additional classes are identified in the plains areas. They are

- ice cap: permanent ice covers more than 50% of the area
- poorly drained lands: lakes or swamps cover more than 10% of the area

Figure 9.5 shows how some of these classes of landscape mosaics are distributed in Köppen's Mediterranean (*Cs*), or subtropical dry summer, zone.

## Effect of Lakes on Zonation

Lakes may have remarkable effects on the surrounding land. One of the best examples is when air masses pass over the Great Lakes in winter. Although cold, the lake water is warm relative to the air, and evaporation supplies moisture to the air mass. Once the air leaves the lakes to pass onto the warmer land on the eastern or southern shore, it becomes unstable and produces copious snowfall (Fig. 9.6).

Lakes also affect the zonal pattern of temperature. Like the margins of the continents, the shore lands have more moderate temperatures than farther inland. For example, the interior of the Michigan peninsula is much colder than the shorelines of Lakes Michigan and Huron. Areas in central Michigan have extreme ranges of more than 20°C compared

**Figure 9.4.** (*Continued*).

**Figure 9.4.** Examples of terrain classes: (**a**) flat plains of eastern Colorado; (**b**) Book-cliffs near Grand Junction, Colorado, part of a well-defined tableland in the Colorado Plateau (photograph by Soil Conservation Service); (**c**) southern Arizona, plains with mountains; (**d**) hills in Pennsylvania, local relief is 150–300 m; (**e**) high mountains in Olympic National Park, Washington, relief is more than 900 m (photograph by Jack Boucher, National Park Service).

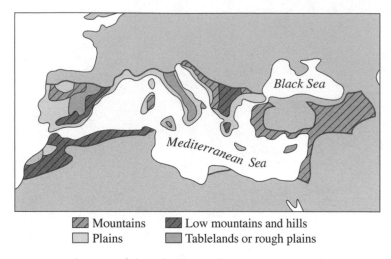

**Figure 9.5.** Landscapes of the subtropical dry zone in the Mediterranean region. From Thrower and Bradbury (1973); redrawn with permission from Springer-Verlag, Heidelberg.

**Figure 9.6.** Mean seasonal snowfall (cm) in vicinity of Lakes Superior and Michigan. From Hidore and Oliver (1993), p. 185. Copyright © 1993 by Macmillan College Publishing Company, Inc.; reprinted with the permission of Simon & Schuster, Inc.

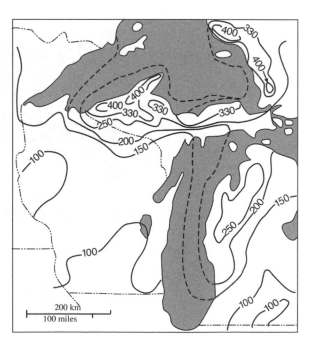

with shorelines with ranges of 18°C. The linkages between lake and land ecosystems are a smaller-scale counterpart to the linkages between the oceans and the continents.

## Effect of Landform on Site Patterns

According to its physiographic nature, a landform unit consists of a certain set of sites. A delta has differing types of ecosystems from those of a moraine landscape next to it. The sites are arranged in specific patterns, according to the way they break up the zonal climate. The mountains and tablelands of the west-central part of North America illustrate this (Fig. 9.7). For example, the high Idaho Mountains and the high-relief tablelands of the Yellowstone Plateau are both located in the Rocky Mountains, a temperate-steppe regime highland. Figure 9.8 shows how these different landforms in the same climate affect site patterns. The Idaho Mountains are made up of various site-specific ecosystems in a complex pattern, including riparian, forest, and grassland. Deep dissection of the mountain range has resulted in variously oriented slopes with varying local climates. Steep slopes oriented at different angles to the sun add complexity to the otherwise simple arrangement of altitudinal zones. Slopes that face toward the sun not only absorb much more heat than those that receive the sun's rays obliquely but also

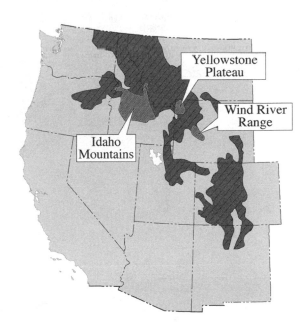

**Figure 9.7.** Selected landform units in the temperate steppe regime mountains of the western United States.

receive many more hours of sunshine. The north sides of east and west valleys are much warmer than the south sides at the same altitude, and these temperature differences are reflected in striking contrasts in vegetation. The Yellowstone Plateau, however, does not have these spotty distribution patterns because its landform is relatively uniform.

We can most accurately delineate units at this level by considering the toposequence (Major 1951), or catena, of site types throughout the unit.

# Geologic Substratum

The geologic substratum is another source of variation within the principal landform classes. Hammond's classification emphasizes the character of the surface form rather than the geologic structure and development history. As such, similar land surfaces may have different underlying rocks. For example, the Fall Line, which separates the Appalachian Piedmont from the Atlantic Coastal Plain on the Fenneman map (1928), appears on the Hammond map as only a few short segments of boundary.

**Figure 9.8.** Landform effects on the montane zone in the temperate steppe regime highlands: (*left*) mountain, Idaho Mountains, Idaho; (*right*) tableland, Yellowstone Plateau, Wyoming. *Left*: sketch by Nancy Maysmith, from photograph. *Right*: photograph by U.S. Forest Service.

Although the Fall Line represents a major break in geologic structure, it forms a poor dividing line for surface configuration. The narrower valley floors, more rolling divides, and higher elevation that distinguish the Piedmont surface from much of the inner Coastal Plain occur only in places distinctive enough to warrant setting apart by a Hammond class boundary. Nevertheless, the ecosystem patterns of these two geologic units are different because of differences in the variety of relief and roughness.

## Effect of Geology on Zonal Boundaries

Geologic factors may modify zonal boundaries. Isachenko (1973) described how this works: In uniform geologic-geomorphological conditions, the transition between adjacent zones is often extremely diffuse. Where, however, the surface is variegated, zonal boundaries assume a more distinct form. Thus, the northern boundary of the forest-steppe zone on the Russian Plain lies along the interface of two distinct types of geology: elevated, dissected plains with loess-type carbonate soils, and

**Figure 9.8.** (Continued).

low-lying sandy forest areas. The former favors the growth of broad-leaved forests and the spread of steppe grasslands. The latter, by contrast, favors a southward shift of the tayga's swamps and conifer forests. Accordingly, the boundary between the forest (tayga) and the forest-steppe zones generally lies directly along the interface of such lithologic regions.

In the Baltic region, owing to the widespread distribution of carbonate rocks, the northern boundary of the mixed-forest zone is displaced far to the north, so that its actual position varies with the theoretical position (Fig. 9.9). In fact, the zonal boundary would lie much farther south if we used the zonal-climatic criteria. Kruckeberg (2002) gives additional examples of this process.

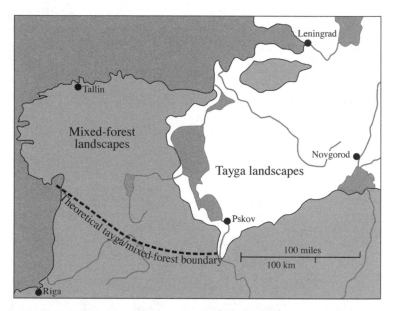

**Figure 9.9.** Boundary between the tayga and mixed-forest zones on the northwestern Russian Plain. From Isachenko (1973), p. 101; reproduced by permission of John S. Massey (ed.).

Geologic structure is an important factor in differentiating mountain landscapes. It is more complex than along the plains. In mountains, the lithology and position of the substrata change more frequently. Unlike the plains, mountains are composed essentially of dense sedimentary and igneous rocks, constituting the immediate substrata for soil formation and plant cover. The effect of bedrock on these other ecosystem components is well known. The soil-forming processes differ on the sedimentary and massive-crystalline rocks. Weathering and soil formation change massive-crystalline rock more than sedimentary rocks. Acid and basic rocks have a different effect on the migration of chemical elements, and associated processes. The podzolizing process fully develops on acidic, crystalline rocks, rich in silicon. On basic rock, soils are rich in humus. Accordingly, the line dividing outcrops of different kinds of rock constitutes an important ecological boundary; on either side of this boundary different landscapes prevail.

The effect of substrata on soils and vegetation is most marked in dry or cold climates. Here, soil development is slow, so that the mineral composition of the parent material often predominates in the thin soil cover. The availability of water in different soils may differ so widely that different vegetation occupies them in the same desert climate. The result is a mosaic of ecosystems unmatched in most humid climates, where soil development and subtle vegetational differences tend to mask the effects of the underlying rocks.

# Levels of Landform Differentiation

We can consider landscape mosaics, or subecoregions, at three levels. Of these the broadest, *sections*, are based on broad land-surface form classes following the system of Hammond (1954, 1964). We determine *subsections* by subdividing sections into areas with homogeneity of lithologic structure, which reflects differentiation at a different scale. In other words, the major landform differences result from the overall shape of the surface. The next scale of landforms reflects differences in lithology within the overall shape. This can be illustrated by considering the Wind River Range in Wyoming (see Fig. 9.7). The area is a high, 4200 m mountain, in the temperate-steppe climatic regime. On the southwest side of the mountains, erosion has exposed a Precambrian granitic core (Fig. 9.10). On the northeast side, exposed Paleozoic and Mesozoic rocks dip steeply to the northeast. These latter rocks, of varying degree of erosivity, form a broad band along the northeast side of the range.

These opposite flanks of the range have vastly different patterns of weathering, erosion, and chemical degradation. Hembree and Rainwater (1961) report that the rate of degradation by solution on the northeast flank is twice that on the southwest flank—about 17 and 9 tons per square kilometer per year, respectively. Conversely, the stream runoff on the southwest flank is about 1.5 times that on the northeast. This seeming anomaly is principally due to the erosive nature of the girdling band of Paleozoic and Mesozoic rock on the northeast flank.

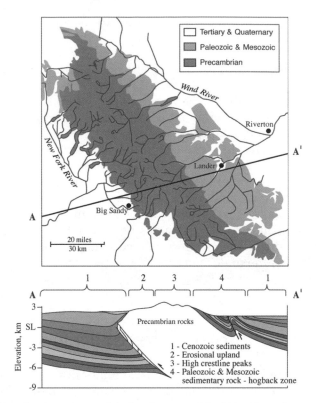

**Figure 9.10.** Geologic base of the landscapes of the Wind River Range, Wyoming. From Love and Christiansen (1985).

Within a uniform geologic base, macrorelief, and climate, there may exist mesorelief features that produce variation in local climate and microclimate. Sites, elsewhere defined as "landtypes" (ECOMAP 1993), are formed within the limits of a single type of mesorelief, uniform bedrock, hydrologic condition, microclimate, and soil. Geographic associations of site units form the most detailed level of a hierarchy of landscape mosaics.

In Chapter 5 and the beginning of this chapter, we discussed macrorelief that includes major continental features of azonal origin (i.e., those produced by tectonic movement and geologic structure) (e.g., plateaus, mountain ranges). Mesorelief is the sculptural variety of those features. It comprises the detail against the background of major features, mainly caused by erosion and deposition (e.g., various erosional, glacial, windborne, and karst forms).

We may regard a uniform geologic structure, together with its set of sculptural variations, as a *geomorphic complex*, or *unit*. Such a complex is commensurate with landscape mosaic. It has a uniform geologic base and is subject to the same geomorphologic processes. Examples of geomorphic complexes are (1) crystalline shields with a complex of glacial erosional and fluvioglacial depositional forms, (2) structural plateaus composed of limestones and dolomites and capped by glacial and karst forms, and (3) inter-mountain tectonic depressions filled with alluvium and other deposits.

The organic world of a landscape mosaic consists of a variegated complex of site-specific ecosystems. By contrast with a site, a landscape is not characterized by any one plant community. A single landscape at the landscape-mosaic level may include plant communities belonging to different types of vegetation (e.g., almost every landscape in the tayga zone includes forest, swamp, and grasslands, and occasionally even tundra). Similarly, a specific plant community may extend over many landscapes.

A similar relationship exists between the landscape mosaics and soils. It is difficult to find landscapes with only a single type of soil. Various soils frequently alternate over a small area, each associated with a single site. Accordingly, a landscape at this level corresponds to a soil association.

Landscapes at this landscape-mosaic level consist of a pattern (catena or association) of local ecosystems (microecosystems) matched to the sequence of topographic facets. Similar facets have similar local climate and hydrologic conditions. Many names have been proposed for these units. Milne, in his classic soil survey of East Africa (1936), proposed the term *physiographic complex* for the association and pattern of soil types in a natural region. He named the sequence of soils encountered between a hill crest and the valley floor a "catena" (Latin for "chain"). The Australians (Christian and Stewart 1968) call them "land systems," the Russians (Isachenko 1973) "landscapes." Wertz and Arnold (1972) use "land type association," and this term has been adopted by ECOMAP (1993). The first example of such a landform-vegetation-soil catena is taken from Rowe and Sheard (1981) and is from the low subarctic ecoregion of the Lockhart River area of Canada.

The region is part of the Precambrian crystalline shield (the Canadian Shield) whose surface is mantled by various kinds of fluvioglacial deposits. These include tracts of bedrock, thin drift, stratified drift (ridged or smooth), moraine (drumlinized, transverse-ridged, or smooth), alluvial, lacustrine, and peat terrain. The forest is open boreal woodland (lichen woodland with bog forest in lowlands), known as tayga. It has a continental subarctic climate.

The sites of this area represent about six types. The ecological interrelationships of the types are demonstrated schematically in Figure 9.11.

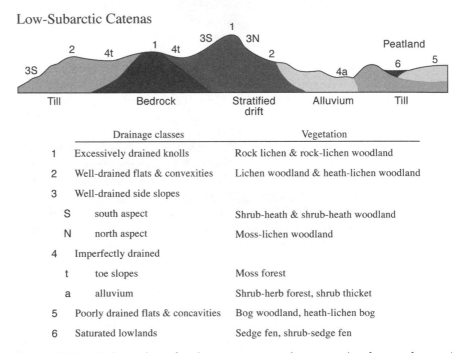

**Figure 9.11.** Relationship of soil, vegetation, and topographic facets of typical landforms in the low subarctic ecoregion of the Lockhart River area of Canada. From Rowe and Sheard (1981).

A second example illustrating the landscape ecosystem is from the montane zone of the Southern Rocky Mountains, a series of massive ranges separated from one another by broad valleys or extensive basins. Although the mountain ranges differ in details among themselves, their common characteristics permit useful generalizations. The altitudinal strata of ecosystems occur within predictable elevation limits throughout the region because they are embedded in the same climate. Different patterns of site-specific ecosystems occur on different relief and geologic structures within the range (see discussion of Wind River Range above). Variation in sculptural forms is superimposed over the altitudinal climatic stratification and geologic structure. Most of the high peaks and plateaus of these mountain ranges have been glaciated. Surrounding them are subdued but deeply dissected uplands, some of which have been subject to cryic, or periglacial, processes. The upper valleys of streams flowing down from the central divide usually show the marks of active valley glaciation. They are deeply incised in narrow, steep-gradient canyons near the margins of the range.

Such differentiations are to be seen in Figure 9.12, a volcanic canyon landscape below the Lower Fall of the Yellowstone River. The hilly land in the background has the normal plant cover of the region, a montane coniferous forest. The canyon in the foreground changes to treeless slopes interspersed with partially tree-covered slopes as well as areas of rock outcrop and talus. By modifying the macroclimate to topoclimate, such distinctive geomorphic units within a vegetation zone support a separate landscape composed of a pattern of sites, or landtypes. This kind of unit is also defined as a *landtype association* (ECOMAP 1993).

**Figure 9.12.** Landscapes in a mountainous region (Yellowstone Plateau). Altitudinal limits are determined by climate, whereas different landscapes within an altitudinal belt are determined by geomorphic and geologic conditions. Photograph by George A. Grant, National Park Service.

Thus we can identify landforms of different scales to establish a hierarchy of landscape mosaics (Table 9.2).

**Table 9.2.** Hierarchy of landscape mosaics

| Section | Land-surface form (e.g. plateaus, mountains) |
|---|---|
| Subsection | Lithology |
| Landtype association | Geomorphic complex, or unit (uniform geologic structure subject to the same geomorphic processes) |

# Landforms in Review

Landforms add another dimension of variation within the broad climatic regions. Geologic processes play an important independent role in landform evolution through tectonic and volcanic activity that has shaped the continents and the major landform units within them. The surfaces of the continents are shaped into a remarkable variety of surface configurations, called landforms. They strongly influence the distribution of ecosystems through modification of the climate and strongly control land use. Inversely, climate influences landforms through the same factors of heat and moisture that control differences in soil and plant cover.

# CHAPTER 10

# Microscale: Edaphic-Topoclimatic Differentiation (Sites)

W̶e may subdivide landscape mosaics into smaller ecosystems called sites or microecosystems. At this point, we turn our attention to the component parts of these mosaics. These are minor in the sense of geographic scale but may play a decisive role in determining the land use.

Although macroclimate and broad-scale landform patterns control the distribution of ecological regions and landscapes, local climate and ground conditions (especially soil moisture availability) control local differences. The latter is the edaphic (related-to-soil) factor. Other things being equal, the edaphic patterns of a landscape will determine the spatial patterning of the biota (Wiens et al. 1985).

## Causes of Site Pattern

Site patterns result from variability in fine-scale landform pattern within relatively uniform landform classes. Figure 10.1 shows how finer-scale landform patterns are formed. Initial landforms are formed by crustal activity. The energy for crustal activity has an internal energy source. As explained in Chapter 5, this heat energy is generated largely by natural radioactivity and primordial heat. Landforms shaped by processes and agents of denudation belong to a group of sequential landforms.

R.G. Bailey, *Ecosystem Geography*, DOI 10.1007/978-0-387-89516-1_10,
© Springer Science+Business Media, LLC 2009

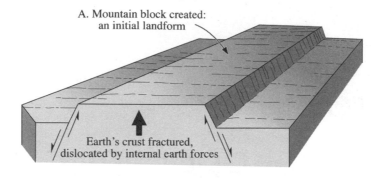

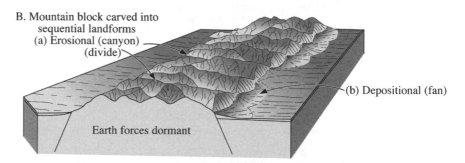

**Figure 10.1.** Initial and sequential landforms. **A**. mountain block uplifted by crustal activity; **B**. uplifted crustal block that has been attacked by agents of denudation and carved up into a larger number of finer-scale landforms. From Strahler and Strahler (1996). From Strahler and Strahler. *Elements of physical geography*, 4E. © 1996 by John Wiley & Sons, Inc.; reprinted with permission.

# Slope-Aspect and Ground Conditions

Within a landform, slight differences exist in slope and aspect that modify the macroclimate (or mesoclimate) to local climate. Geiger (1965) used the term "microclimate" for the climate at or near the ground surface, such as within the vegetation or soil layer. Microclimate directly influences ecological processes and reflects subtle changes in ecosystem function and landscape structure (Chen et al. 1999; Swanson et al. 1988). Thornthwaite (1954) referred to these modifications of climate as topoclimate (i.e., the climate of a small place). The three commonly identified classes of topoclimates, based on temperature, are normal, hotter than normal, and colder than normal (Fig. 10.2). These topoclimates are subdivisions of the macro- and mesoclimates (Fig. 10.3). We refer to the ecosystems controlled, and partially defined, by topoclimate as *site classes*, following Hills (1952).

**Figure 10.2.** Topoclimates: effects of slope and aspect on temperature.

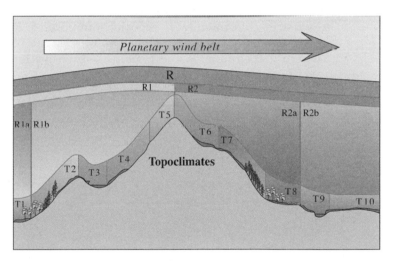

**Figure 10.3.** Topoclimates in relation to the higher levels of climatic division. From Yoshino (1975) in Barry (1992), p. 12; reprinted by permission of Routledge.

When we differentiate local sites within topoclimates, soil-moisture regimes provide the most significant segregation of the plant community. A sequence of moisture regimes, ranging from drier to wetter from the top to the bottom of a slope (Fig. 10.4), may be referred to as a soil catena, or a toposequence (Major 1951). Exposure to wind also influences soil moisture. The existence of small relief forms substantially affects the movement of air masses; it changes the direction and velocity of winds near

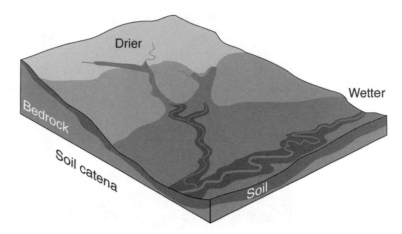

**Figure 10.4.** Toposequence or catena of soil moisture regimes.

the ground, thus contributing to the redistribution of rainfall. The wind-ward hill slopes usually receive less rain than the lee slopes (Fig. 10.5). Redistribution of snow is especially important. From the hilltops, and often from the windward slopes as well, snow blows into depressions where it accumulates and remains 1–2 weeks longer than on elevated sections.

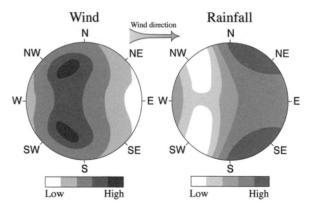

**Figure 10.5.** Distribution of wind velocities and rainfall around a hill. From Geiger in Isachenko (1973), p. 140; reproduced by permission of John S. Massey (ed.).

The effect of topoclimate on different surface materials and vege-tal structures is determined to some extent by their different albedos, because these determine the amounts of solar heat they absorb. In gen-eral, the darker the color of the soil or rock and the more complete the plant canopy, the lower the albedo. Albedos range between extremes of

95% for new snow surfaces and less than 6% for water (Sellers 1965). Most albedos of most soil and vegetation lie between 10 and 40%.

A common division of the soil moisture gradient is dry, fresh, moist, wet, and very wet. Table 10.1 relates the most common type of soil associated with these and other categories. Production of tree species in the northeast part of North America is related to this gradient (Fig. 10.6). The influence of moisture on the local distribution of plant communities is well illustrated by the grassland vegetation of the temperate steppes. In a

**Table 10.1.** Humidity of ecosystems (adapted in part from Crowley)

| Humidity category | Most common soil type |
| --- | --- |
| Aquatic | Water |
| Wet | Bog, marsh, swamp, tidal marsh, very poor drainage |
| Very humid | Gleysols or low floodplain, poor drainage |
| Moist | Gleyed soils, imperfect drainage; or slopes protected from sun |
| Fresh | Soils of medium texture, good drainage |
| Slightly dry | Shallow or sandy soils, excessive drainage; or slope exposed to sun |
| Dry | Sand, sandy Regosols, Lithosols |
| Very dry | Gravel, very shallow Lithosols |
| Xeric | Outcrops not fed by seepage, deep deposits of rocks |

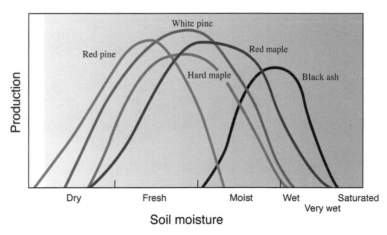

**Figure 10.6.** Production gradients of tree species relative to soil moisture in northeastern North America. From Hills (1976), p. 79.

study of prairie, meadow, and marsh vegetation in Nelson County, North
Dakota, Dix and Smeins (1967) (as reported by Smith 1977) divided the
soils into ten drainage classes, ranging from excessively drained to permanently standing water. They determined the indicator species for each
drainage class and then divided the vegetational display into six corresponding units (Fig. 10.7). The uplands fell into high prairie, midprairie,
and low prairie and the lowlands into meadow, marsh, and cultivated
depressions. High prairies dominated the excessively drained areas and
were characterized by needle-and-thread grass (*Stipa comata*), western
wheatgrass (*Agropyron*), and prairie sandreed (*Calamovilfa longifolia*).
The midprairie, considered to be the climax or true prairie, was dominated by big bluestem (*Andropogon gerardii*) and little bluestem, porcu-
pine grass (*Hesperostipa spartea*), and prairie dropseed (*Sporobolus het-
erolepis*). Low prairie on soils of moderate moisture was characterized by
big bluestem, little bluestem, yellow Indian grass (*Sorghastrum nutans*),
and muhly (*Muhlenbergian* spp.). Lowlands that occupied soils in which
drainage was sluggish and the water table was within the rooting depth
of most plants were characterized by canary grass (*Phalaris* spp.), sedge
(*Carex*), and *Scolochloa festucacea*. Meadows on even wetter soils were
dominated by northern reedgrass, wooly sedge, and spikerush. Marshes

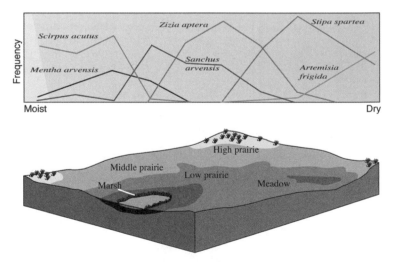

**Figure 10.7.** Prairie vegetation forms a mosaic that is influenced by topography
and drainage regimes. (*Bottom*) A hypothetical block diagram of a North Dakota landscape showing the relative positions of vegetation units; (*top*) distributions of selected
species along a drainage gradient. From Dix and Smeins (1967) in Smith (1977), p.
124. From Smith. *Elements of ecology and field biology* © 1977 by Robert Leo Smith;
reprinted with permission of Pearson Education Inc.

that contained permanently standing water contained stands of reed, cattails, and Tule bulrush.

For an area of shortgrass steppe in Colorado, Dodd et al. (2002) found that the dominance of woody plants is associated with coarse textured soils, and that ecotones between woody and herbaceous plant functional types are associated with soil textural changes. Likewise, Walker (2000) found that vegetation patterns in the Arctic tundra are determined by variation in landform factors, primarily through their effects on soil moisture and snow regimes. In the till plain of the Eastern Broadleaf Forest of Indiana, Dolan and Parker (2005) emphasized landform and landform component, which are related to the swell-and-swale topography of the till plain, as the most important factors in determining differences between plant community composition.

# Geologic Differentiation

The physical character of the bedrock also affects vegetation patterns. Different kinds of rock vary in their resistance to erosion, their hydrologic properties (porosity, permeability, and so on), and chemistry. This

**Figure 10.8.** Lithosequence of vegetation in Canyonlands National Park, Utah. Photograph by Richard Frear, National Park Service.

affects not only the topography, but also soil formation and subsequent moisture content. This is particularly well illustrated in the semiarid regions with sedimentary rock, such as the Colorado Plateau (Fig. 10.8). Here, the bedrock is interbedded sandstone and shale. The shale erodes more easily, forming soils with higher moisture. Such soils support a more dense vegetation consisting of lightly but scattered grasses, shrubs, and small trees. This banding, a lithosequence, is caused by preferences of vegetation for the greater moisture of slopes underlain by rock with slightly greater moisture. However, we can still account for these variations within our classification scheme because they relate primarily to variations in soil moisture.

# Topoclimate-Soil Moisture Ecoclimatic Grid

Topography, even in areas of uniform macroclimate, leads to deviations from normal topoclimate and mesic soil moisture (Fig. 10.9). We can use a simple three-by-three grid (Fig. 10.10) and characterize any region. Two factors comprise the grid, namely, ecoclimatic regime (i.e., local climate as influenced by local topographic position) and soil-moisture regime (as influenced by topographic position and soil materials). The climatic climax (see discussion of climax in Chapter 5) theoretically would occur over the whole region, except for topography leading to different local climates. In other words, if the region were relatively flat and covered by soils with mesic, soil-moisture regimes, then all sites would be described by the central cell of the matrix. Because that is rarely the case, other possibilities exist. Sites resulting from variation in soil moisture and temperature and portrayed by the other matrix cells are referred to as edaphic climaxes.

**Figure 10.9.**
Deviations from normal topoclimate and soil moisture within an ecoclimatic zone.

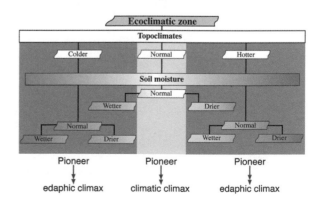

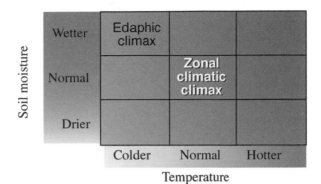

**Figure 10.10.** A matrix of nine sites that provide the basis for characterization of vegetation-site relationship in a region.

**Table 10.2.** Site types

| Climatic climax | Zonal |
|---|---|
| Edaphic climax | Azonal |
| | Intrazonal |

These deviations occur in various combinations within a region and are referred to as *site types* (Hills 1952). As a result, every regional system—regardless of size or rank—is characterized by the association of three types of local ecosystems or site types (Table 10.2): zonal, azonal, and intrazonal.

## Zonal Site Types

Normal topoclimate and fresh and moist soil moisture characterize these sites (e.g., the sagebrush terraces in Jackson Hole, Wyoming) (Fig. 10.12). The lowland climate here is semiarid, and the climax vegetation here is normally sagebrush semidesert.

## Azonal Site Types

These sites are zonal (they occupy normal environments) in a neighboring zone but are confined to an extrazonal environment in a given zone. For instance, in the Northern Hemisphere, south-facing slopes receive more solar radiation than north-facing slopes, and thus south-

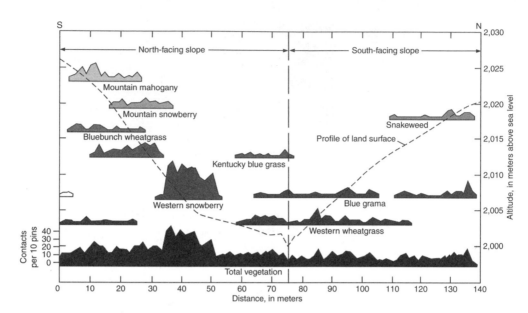

**Figure 10.11.** Distribution of selected plant species on north- and south-facing slopes on Green Mountain near Denver, Colorado. From Branson and Shown (1990).

facing slopes tend to be warmer, drier, less thickly vegetated, and covered by thinner soils than north-facing slopes. In arid mountains, the south-facing slopes are commonly covered by grass, whereas steeper north-facing slopes are forested (see Fig. 1.11). Contrasts are notable features of most landscapes with moderate relief as well. Branson and Shown (1990), in their study on north- and south-facing slopes in the Denver area, found marked contrast not only in the foothills themselves but also along gentler slope on the High Plains, where shrubs trace precisely the north-facing slopes (Fig. 10.11). Likewise, south-facing slopes are notable for dotted patterns of widely spaced plants or significant species changes between northern and southern exposures. Azonal sites are hotter, colder, wetter, and drier than zonal sites. The riparian forest growing on wet sites adjacent to the Snake River (Fig. 10.12) is a good example. This forest, which is growing in a semiarid climate, is doing so only because of the presence of the high water table—not because of the climate. Another example is where rocky reservoirs support ponderosa pine [also called rock pine by early botanists; as reported by Woodward (2000)] within grasslands of the Great Plains.

The size, direction, and configuration of valleys and basins are also important in determining azonal conditions. For example, valleys and basins have quite different diurnal thermal regimes from their surrounding slopes. During the course of a day, the air in the bottom of depres-

**Figure 10.12.** View of Jackson Hole, Wyoming, from the east. In the foreground is the broad valley of the Snake River, with terraces above; Teton Range (altitude to about 3700 m) in background. Photograph by National Park Service.

sions tends to become warmer, and air currents tend to move up the side slopes (Fig. 10.13a). At night, this condition is reversed, and cool air, being heavier, tends to move downslope into depressions and form cold air or frost pockets in which mist or fog often occurs.

When the hollow is an elongated valley rather than a closed depression, air movement up and down the valley occurs (Fig. 10.13b). In the daytime, after the warmed air has begun to rise up the valley slopes, a second movement of air takes place, up the valley itself. Similarly, at night air flows down the valley.

The downward movement of air at night is not a phenomenon confined to depressions but also occurs on slopes of isolated hills. We can discern a twofold subdivision of topographic situations according to diurnal temperature variations. In response to night cooling in a valley, a "lake" of cold air is located near the bottom of a valley. A zone of higher temperatures, known as a thermal belt, develops on the slopes.

The contrast between warm slopes and cold valleys is so great that some valleys may experience temperatures many degrees colder than mountain stations thousands of meters higher. According to Miller (1946)

Early morning                    Night

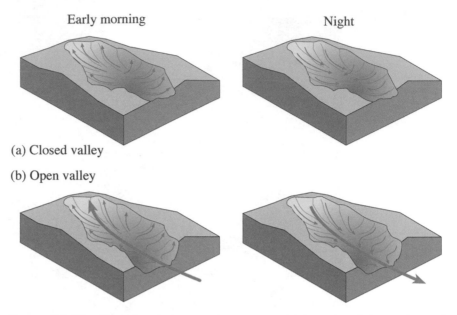

(a) Closed valley

(b) Open valley

**Figure 10.13.** Diurnal variations in air movement (**a**) in a closed depression and (**b**) in a valley. From Geiger (1965) as modified in Mitchell (1973), p. 101.

(as reported by Mitchell 1973), the lowest temperature on record in the United States (–54°C) was recorded at Miles City, Montana, lying in a deep depression in the Great Plains, whereas Pike's Peak, which is 3400 m higher, has never recorded a temperature below –40°C.

Vegetation reflects these air movements. Cold-air drainage (the cold downdrafts) in the montane zone creates grassy areas in the valleys that are too cold for tree growth. Early settlers referred to these treeless areas as "parks" (Fig. 10.14).

These air movements also affect the high-altitude timberline in non-tropical mountains (Troll 1968). Valleys and gullies that descend from the alpine zone tend to be treeless because cold air accumulates there in winter and leads to temperature inversions. Thus forests can grow higher on the adjacent intermediate slopes than in the valleys.

## Intrazonal Site Types

These sites occur in exceptional situations within a zone. They are represented by small areas with extreme types of soil and intrazonal vegetation. Soil influences vegetation to a greater extent than climate, and thus the same vegetation forms may occur on similar soil in several zones. We can differentiate them into *five* groups:

**Figure 10.14.** Park in the Santa Fe National Forest, New Mexico. Photograph by Bluford W. Muir, USDA Forest Service.

1. The first site type is *unbalanced chemically*. Some examples from the United States are the specialized plant stands on serpentine (magnesium-rich) soils in the California Coast Ranges. Other examples are the belts of grassland on the lime-rich black belts of Alabama, Mississippi, and Texas (Fig. 10.15) and the low mat saltbush (*Atriplex corrugata*) on shale deserts of the Utah desert, which contrasts with upright shrubs on adjacent sandy ground. The kind and amount of dissolved matter in groundwater also affect plant distribution. This is especially obvious along the coasts and along edges of desert basins (Fig. 10.16) where the water is brackish or saline. Plants adapted to moist saline ground are called halophytes.

2. *Very wet* sites occur where the groundwater table controls intrazonal plant distributions. The plants of these sites are phreatophytes, plants that send roots to the water table. Examples include riparian zones in the deserts of the southwestern United States, such as a cottonwood (*Populus deltoides*) floodplain forest and the cypress (*Taxodium distichum*) and tupelo (*Nyssa aquatica*) forests of the Southeast (Fig. 10.17).

**Figure 10.15.** Intrazonal site types on extreme types of soil. Limy formations in Alabama support grasses in the midst of southern pine forest growing on the less limy formations. From Hunt (1974), p. 170.

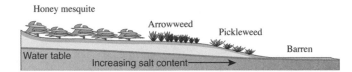

**Figure 10.16.** An example of plant distribution controlled by salts dissolved in ground-water at the edge of a salt pan in Death Valley, California. From Hunt (1974), p. 172.

**Figure 10.17.** Louisiana swamp cypress. Photograph by Clement Mesavage, U.S. Forest Service.

3. *Very dry* sites with sandy soils, because of limited moisture-holding capacity, are drier than the general climate. At the extreme, sand dunes fail to support any vegetation because they are too dry (Fig. 10.18).

4. Another site type is *very shallow*. Soil depth, as a factor in plant distribution, may be controlled by depth to a water table or depth to bedrock. Vegetation growing along a stream or pond differs from that growing some distance away where the depth to the water table is greater. Examples of the influence of depth to bedrock on plant distribution can be seen in mountainous areas where bare rock surfaces that support only lichens are surrounded by distinctive flowering plants growing where thin soil overlaps the rock and is, in turn, surrounded by forest where the soil deepens (Fig. 10.19).

5. *Very unstable* sites are areas where gravity combined with high relief, steep slopes, weak bedrock, excessive groundwater, earthquake shocks, and undercutting causes landslides. These slides include slump earthflows, rockslides, rockfalls, mantle slides, avalances, and mudflows. Commonly, these slides produce vegetation anomalies. For example, in the Middle Rocky Mountains on slopes between 1800-

**Figure 10.18.** Sand dunes in Death Valley, California. Photograph by George Grant, National Park Service.

**Figure 10.19.** Spotty vegetation pattern caused by shallow soil in the alpine zone of the Beartooth Mountains in Yellowstone National Park, Wyoming. Photograph by George Grant, National Park Service.

and 3200-m elevation, dense aspen (*Populus tremuloides*) growth, in an area normally supporting evergreen forest, indicates the location of earthflows (Fig. 10.20). Earthflows do not support slow-growing conifers; instead because of soil movement, fast-growing aspen replace them, indicating wet ground conditions. Aspen may be present because they reproduce mainly by root suckers. Soil movement disturbs the roots, stimulating sprouting and probably contributes to its spreading. Continued movement shifts and tilts the trees.

## Examples

Southern Ontario, Canada.    Figure 10.21, in a simplified way, illustrates how topography, even in areas of uniform macroclimate, leads to differences in local climates and soil conditions. This example is from southern Ontario, Canada (Hills 1952). On level or moderately rolling areas where the soil is well drained but moist, a maple-beech community (*Acer-Fagus*) (sugar maple and beech being the dominant plants) is the terminal succession. Because we find this type of community repeatedly

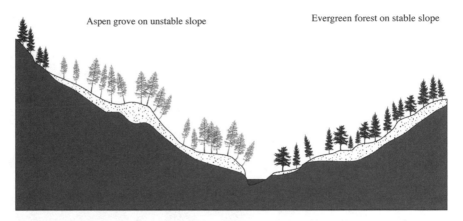

**Figure 10.20.** Vegetation anomalies reveal active landslides.

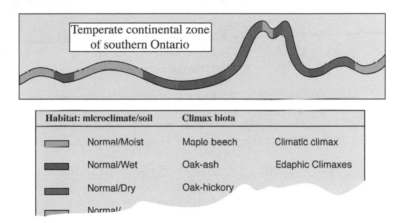

**Figure 10.21.** Different forest climaxes occurring in the temperate continental zone of southern Ontario, Canada. Simplified from Hills (1952) in Odum (1971), p. 265. From Odum. *Fundamentals of Ecology*, 3E. © 1971 Brooks/Cole, a part of Cengage Learning, Inc. Reproduced with permission. www.cengage.com/permissions (Diagram is truncated; only three of nine possible environments displayed).

in regions wherever land configuration and drainage are moderate, the maple-beech community is judged to be the normal unmodified climax of the region. Where the soil remains wetter or drier than normal, a somewhat different end-community occurs, as indicated. The climatic climax theoretically would occur over the entire region except for topography leading to different local climates, which partially determines edaphic conditions. On these areas, different edaphic climaxes occur; climatic climaxes occur only on mesic soils.

The conditions illustrated also extend into the northeastern United States from the Great Lakes to New England. A map of the original vegetation of a portion of Monroe County, New York, showing climax and edaphic climaxes is included as Figure 10.22.

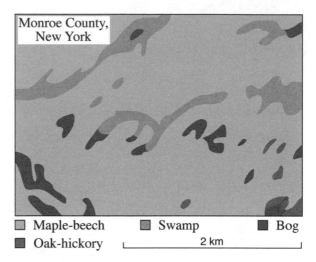

**Figure 10.22.** An example of microscale ecosystem patterns. The beech-maple forest, which is the climax, would occur over the entire area were it not for topography-producing edaphic climaxes (the other vegetation communities shown). From Shanks in de Laubenfels (1970), p. 75; reproduced by permission of the author.

The units at this scale correspond to units with similar soil particle size, mineralogical classes, moisture, and temperature regimes. These are generally the same differentiating criteria used to define families of soils in the System of Soil Taxonomy of the National Cooperative Soil Survey (USDA Soil Conservation Service 1975).

The potential, or climax, vegetation of these units is the plant community with the rank of plant association, which is the basic unit of phytocenology (cf. Table 3.1). Associations (also called habitat types in the western United States by Pfister and Arno [1980]) are named after the dominant species of the overstory and the understory (Daubenmire 1968) (e.g., grand fir/ginger).

The use of the word *potential* is critical because it allows a single site to include different kinds of vegetation as long as they represent different stages of biotic succession from weedy pioneers to "climax" forest or grasslands. We can identify another level (provisionally called the *site phase*) to allow the classification to communicate the ages and species composition of existing vegetation. These correspond to forest and range-cover types that are commonly mapped by using remote-sensing imagery.

Colorado Front Range.        A second example is taken from the Front Range area of Colorado. The area lies within the temperate steppe zone (see Fig. 5.25). The area is a mountain range with altitudinal belts ranging from dry steppe, to coniferous forest, to mountain vegetation above treeline (Fig. 10.23).

Within this forest cover, the main environmental contrasts in the types of vegetation are not simply related to elevation but to a combination of elevation and topography. We may locate the main forest types on an elevation-topographic gradient (Fig. 10.24). The different types of sites are ordered according to the driest to the wettest conditions. Exposed ridges mark the dry end of the gradient, whereas the wetter end consists of deep ravines with flowing streams. Between these two extremes, other sites are arrayed according to their moisture characteristics. By knowing the elevation and exposure, we can predict the kind of vegetation that is likely to occur there. For example, limber pine (*Pinus flexilus*) forests occur in dry sites at elevations greater than 2600 m, whereas ponderosa pine–fir (*Pinus ponderosa–*

**Figure 10.23.** Example of edaphic-topoclimatic differentiation. Montane zone of Colorado Front Range in Rocky Mountain National Park is broken into a complex mosaic of vegetation types due to local differences in aspect and exposure (see Fig. 10.24). Photograph by National Park Service.

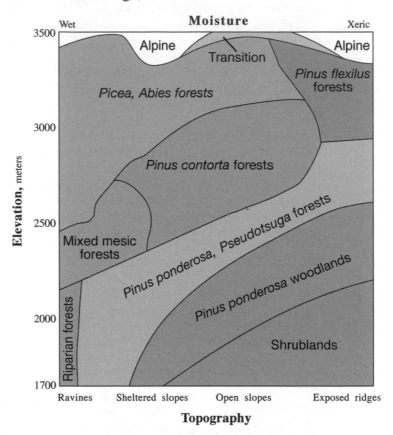

**Figure 10.24.** Boundaries between vegetation types in the Front Range, Colorado. These boundaries are related to two environmental gradients: elevation and exposure. Different types of terrain have been arranged in sequence along the exposure axis, from protected and wet sites to open and dry sites. From Peet (1981), p. 36. Reprinted by permission of Kluwer Academic Publishers.

*Pseudotsuga*) forests occur on all sites, except wet, below this elevation. Two effects cause the difference: the exposure to wind and to solar radiation. On the lee side of the ridges, the wintery snow cover is thicker and lasts longer than on the windward side. Various slope aspects will cause further differences in duration of snow cover, because of the different annual and diurnal amounts of sunshine and shade.

Gregg (1964) sketched a macroscopic view (Fig. 10.25) of the Front Range in Boulder County that shows how the physical environment helps shape the distribution of vegetation types.

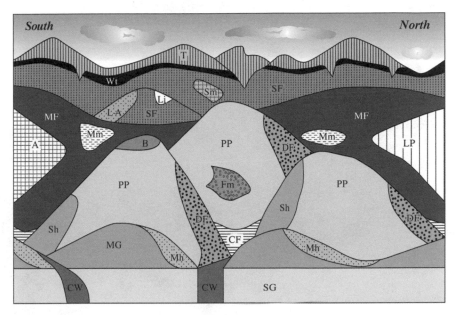

**Figure 10.25.** Diagrammatic distribution of vegetation types in the mountains of the Front Range in Boulder County, Colorado. Note: SG = shortgrass prairie, CW = cottonwood-willow forest, MG = mixed grass prairie, PP = ponderosa pine forest, Sh = mixed shrubs, grass, and yucca, Mh = mountain mahogany brush, CF = canyon forest, Fm= foothills meadow, DF = Douglas-fir forest, MF = mixed montane forest, A = aspen forest, LP = lodgepole pine forest, Mm = montane meadow, B = rocky bald, SF = spruce-fir forest, L-A = lodgepole pine-aspen forest, Li = limber pine forest, Sm = subalpine meadow, Wt = wind timber (krummholz), T = tundra, At = avalanche track. From Gregg (1964).

# Human Influences on Ecosystems and Present-Day Systems

Human influences on ecosystems are readily apparent. The suburban sprawl of Los Angeles into the San Fernando Valley, the falling water levels in the West, and the fouling of the Great Lakes with pollutants are examples. In addition, large areas of the original vegetation have been changed (Fig. 10.26). As a result, many natural ecosystems have become nearly extinct (Noss et al. 1995). Here we may cite the climax forest of the eastern United States, which has been for decades, and will probably continue to be, an agricultural savanna (Fig. 10.27). In other areas, industrial farming of animal products is capable of achieving such high

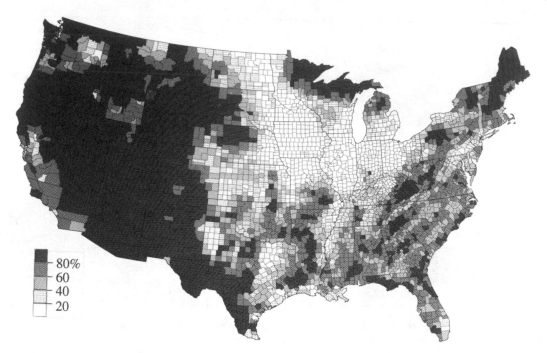

**Figure 10.26.** Percentage of area potentially covered by natural vegetation for counties in the United States. From Klopatek et al. (1979).

yields per acre that the total demand for farmland may be reduced. The increase in woodland, as previously farmed lots are reforested, is related to this trend.

Occasionally, forest plantations are introduced in areas where natural forests have been felled. In grassland areas, both elimination of species and the creation of new hybrids have altered biotic communities. The boundaries between land and water have also been modified by draining wetlands and damming rivers. Changing fire frequency has altered boundaries between plant communities as well. The long-term impact of periodic fire is difficult to judge, but there is some indication that species native to chaparral of the summer-dry Mediterranean zones and the savanna of the winter-dry subtropical zones evolved in association with fire. Exclusion of fires to prevent forest and range fires may cause changes in the composition and density of the vegetation, sometimes with disastrous consequences when fuels build up. Our suppression of wildfires has extended the intervals between major fire events. These efforts have resulted in fires such as the infamous Yellowstone National Park fire in 1988. This fire not only had a different character from past natural fires but it also burned a far larger area.

**Figure 10.27.** Man-modified ecosystem, strip crops in the Eastern Deciduous Forest of Pennsylvania. Photograph by J.W. Wise, Soil Conservation Service.

Despite the low percentage of the earth's surface occupied by cities, these high-density patches have a profound effect on the ecosystem. In the downtown area, the cover of impermeable concrete and asphalt is almost total, which changes the hydrology of the area. Cities drastically alter the climate by the production of heat, the alteration of the surface configuration and its roughness, and the modification of the atmospheric composition. These effects are no longer restricted to the lowlands as urbanization moves increasingly into the mountain ecosystems, such as in the Lake Tahoe Basin and along the Colorado Front Range.

Knowledge about the degree of human-induced modification of ecosystems, or their status, is essential to their conservation. It is important to divide all contemporary ecosystems into several classes, depending on the degree and the character of changes introduced by mankind's activity. Several proposals exist in this area. On a global level, the system developed by the Faculty of Geography at Moscow State University (Milanova and Kushlin 1993) reflects the degree of transformation of present-day landscapes (ecosystems). This notable example defines these categories as

1. *Modal (essentially unaffected) landscapes* have practically no vegetation transformation, and the intensity of human impact is low or
   virtually absent (e.g., ice deserts, high mountain regions, boreal forest, and tundra).
2. *Derivative (secondary) landscapes* are defined as emerging in place
   of modal ones as a result of some human activity (or on previously
   cultivated abandoned lands) but existing in a relatively steady state.
   Among such landscapes are certain Mediterranean landscapes, open
   woodlands of the humid tropics, and deciduous forests of the tayga
   zone.
3. *Landscape anthropogenic modifications* are landscapes where the
   natural components have been more or less changed through inten-
   tional anthropogenic impact. Milanova and Kushlin (1993) divide
   these landscapes into three broad categories: agricultural, silvicultural, and recreational.
4. *Landscape technogeneous complexes* are landscapes where the
   dynamics, environmental status, and socioeconomic functions are
   almost totally determined and controlled by conscious anthropogenic
   impact. Among such landscapes are major water control projects,
   industrial complexes, mining areas, and settlement.

The landscapes are displayed on the *World Map of Present-Day Landscapes* at a scale of 1:15,000,000.

In the United States, several attempts have been made at land-use classification and mapping. From maps of state land use, Marschner (1950)
compiled a map of major land uses at 1:5,000,000. The Soil Conservation
Service (Austin 1965) has mapped land-use regions. Land-use and land-
cover maps are currently being published by the U.S. Geological Survey
(Anderson et al. 1976) at scales of 1:250,000 or 1:100,000. We may also
consider other kinds of maps derived from remote sensing (e.g., forest
types [Powell et al. 1993]; seasonal land cover [Loveland et al. 1991]) as
maps of ecosystem status.

This concludes our review of factors useful in recognizing and mapping ecosystems at various scales. Ecosystem-based planning and development certainly will be all the more successful the more we know about
the ecological differentiation of a territory. We next discuss the applications of this knowledge.

# CHAPTER 11

# Applications of Ecosystem Geography

Ecosytems come in many scales or relative sizes. As I have said, because of the links between systems, a modification of one system may affect the operation of surrounding systems. Furthermore, how a system will respond to management is partially determined by relationships with surrounding systems. Multiscale analysis of ecosystems pertains to all kinds of land. Many planning issues transcend ownership and administrative boundaries. To address these issues, the planner must consider how related ecosystems are linked to form larger systems. *For this analysis to be effective, it must be conducted regardless of ownership or administrative boundaries.* This is because we can only understand ecosystems, as spatial systems, by looking at the whole—not just certain parts. Note that this refers to analysis only, not to government's making management decisions about private land or vice versa.

## Determining the Mapping Units

### How Differently Would We Operate?

We need to map according to permanent features that affect geographic patterns in ecosystems and process, namely climate and landform. The soil and biota that are to be managed result from these controlling factors. When we analyze controlling factors, we can better define permanent boundaries. We would then conduct functional inventories, such as timber, and analyze according to such units. Various disciplines need a common ecosystem unit. This in turn would provide a common unit for integrated management. At present, various disciplines use different

R.G. Bailey, *Ecosystem Geography*, DOI 10.1007/978-0-387-89516-1_11,
© Springer Science+Business Media, LLC 2009

ecosystem units (e.g., stands of trees for foresters and watersheds for water-quality analysts). Analyzing resource interaction is difficult because each discipline selects its own unit of land for analysis. Furthermore, comparing information across disciplinary and administrative unit boundaries is difficult. Under the common ecosystem-unit approach, the various disciplines would all relate to the same area. They would collect and analyze data about a common ecosystem unit. The anthropologist, for example, could define and categorize the status of the system (how human activity has modified the system).

## Sites Are Seen Within the Context of the Larger System

Further ecosystem analysis must be carried out at multiple levels: locally (site) and groups of geographically related ecosites (landscape mosaics and ecoregions).

This approach allows us to see the connection between action at one scale and effects at another. As I said in Chapter 2, landscape mosaics, for instance, delimit areas that represent different patterns or combinations of sites within a regional ecosystem, or ecoregion. When we understand the interaction between sites, processes emerge that were not evident at the site level. The processes of a landscape mosaic are more than those of its separate ecosystems because the mosaic internalizes exchanges among component parts. As discussed in Chapter 1, for example, a snow-forest landscape (Fig. 11.1) includes dark conifers that convert solar energy into sensible heat that moves to the snow cover and melts it faster than in either a wholly snow-covered or a wholly forested basin. The conifers are the intermediaries that speed up the process and affect the timing of the water runoff. Landscapes function differently as a whole than would have been predicted by analysis of the individual elements (cf. Marston 2006). Understanding landscape processes makes it possible to analyze the effects of managing a site on surrounding sites. We can then assess the cumulative effects that may occur from a proposed activity. Without this understanding, the analyst may conduct a hydrologic analysis of the forest and snow-covered areas separately and then add the result to erroneously obtain the total runoff for the landscape. The statement "a system is greater than the sum of its parts" aptly applies to the landscape as well as individual small ecosystems.

**Figure 11.1.** Snow forest landscape in La Sal Mountains, Utah. Sketch by Susan Strawn, from photograph.

# Relationships

### Ecosystem Units and Functional Inventories

How do we determine the relationship between ecosystem units and functional inventories? We would first identify an ecosystem site unit according to climate and landform; second, we would identify existing vegetation community and then make functional inventories as needed (Fig. 11.2). For example, we would classify and map a site as a potential vegetation community of ponderosa-Idaho fescue/soil series. However, the site has been burned and heavily grazed, so the existing community is a cheatgrass-Idaho fescue. Functional inventories would then be completed as necessary. In this example, the site could be interpreted for timber and range management, wildlife habitat, and so forth. It is essential that all functional inventories refer to the same area or aggregation of areas.

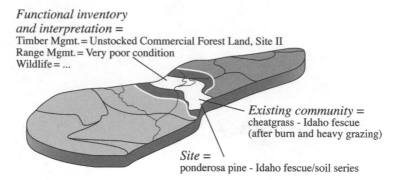

*Functional inventory
and interpretation =*
Timber Mgmt. = Unstocked Commercial Forest Land, Site II
Range Mgmt. = Very poor condition
Wildlife = ...

*Existing community =*
cheatgrass - Idaho fescue
(after burn and heavy grazing)

*Site =*
ponderosa pine - Idaho fescue/soil series

**Figure 11.2.** Relationship between ecosystem units and functional inventories.

## Ecosystem Units and Hydrologic Units

To complete the analysis, we would analyze ecosystems according to the watershed where they reside. A map of watersheds, when overlaid on an ecosystem map, will provide information to predict what effects the alteration of an upstream ecosystem might have on a downstream ecosystem, and visa versa. Conversely, the ecosystem map will provide useful information for those responsible for management of a watershed by showing the kinds of aquatic environments within a watershed. The delineation of watersheds is a useful tool but is not suitable as the primary definition of an ecosystem.

# Examples of Useful Correlations and Applications

## Sampling Networks for Monitoring

Where should we locate monitoring sites? We would like to have detailed information on all the ecosystems involved in a given study area, but that is not feasible because of time and cost. So we must form a sampling strategy that guarantees us, as much as possible, representative information. Two examples of appropriate sampling strategies follow.

Estimation of Ecosystem Productivity.    Land management deals with productivity systems (i.e., ecosystems) from which it attempts to efficiently, and continuously, extract a renewable product, such as wood or water.

We need estimates of ecosystem productivity to assess and manage. To make such estimates, we must develop the relationships between the ecosystem and the information that we need for production. These relationships may be understood at many levels, from simple judgments based on experience, to multivariate-regression and other complex mathematical models. Application of these models is based on the concept of transfer by analogy (i.e., we can apply the information gained on monitored production sites to analogous areas). The analogous areas are ecosystems that have been carefully defined and classified.

Such methods are based on the hypothesis that all replications of a particular class of ecosystem will have fairly similar productivity. Some workers have questioned this hypothesis (e.g., Gersmehl et al. 1982) on the basis that correlations between ecosystem types and behavior are generally low.

The reason for this low correlation is that the criteria used to classify the ecosystem types were applied uniformly over an area without considering compensating factors. These factors may produce the same ecosystem type but for different reasons. For example, soil factors may modify the apparent effects of climate. We know that moisture-demanding plant species often extend into less humid regions on areas of sandy soils because they tend to contain a greater volume of available moisture than do heavier soils. In humid climates, the same soil types support vegetation that is less demanding of moisture than it would be in dry climates. As seen in the following section, it is unlikely that the behavior of a given type of vegetation would be similar in diverse climates.

One way to establish reliable ecosystem-behavior relationships is to divide the landscape into "relatively homogeneous" geographic regions where similar ecosystems have developed on sites having similar properties. For example, similar sites (i.e., those having the same landform, slope, parent material, and drainage characteristics) may be found in several climatic regions. Within a region, these sites will support the same vegetation communities, but in other regions, vegetation on the sites will differ. Thus, beach ridges in the tundra climatic region support low-growing shrubs and forbs, whereas beaches in the subarctic region usually have dense growth of black spruce (*Pinus mariana*) or jack pine (*Pinus banksiana*). Soils display similar trends, as the kind and development of soil properties vary from region to region on similar sites. These climatically defined regions suggest over what areas we can expect to find the same (physiognomically if not taxonomically) kinds of vegetation and soil associations on similar sites (see Hills [1960b] and Burger [1976] for a discussion of regional differences in ecosystem and site relationships).

Theoretically, the influence of climate on the ecogeographic relationships of a region creates unity overall. Such climatic regions delimit patterns of associated aquatic and terrestrial microecosystems over large areas, creating ecosystem regions (see Fig. 10.21). Monitoring the behavior of representative sites makes it possible to predict effects at unmonitored sites within the same ecoregion (Bailey 1991). We must carefully select the monitoring sites to truly represent the region. To do so, they must be drawn from all the types of sites found in a region. Identification of sites based on ecoregional classification could be used to impute their characteristics from sampled sites, for example, using *k*-Nearest Neighbors or similar techniques (McRoberts et al. 2002).

Monitoring sites that represent the kinds of ecosystems found in a region will provide more useful information than those selected otherwise. Data obtained from a representative site will be useful for generalizing and applying to unmonitored sites, thereby lowering the cost and time involved in monitoring.

In recent years, many publications on ecosystems have appeared. However, only rarely (e.g., Breymeyer 1981; Robertson and Wilson 1985) have they used existing information about the geographic variability of ecosystems to design monitoring programs.

Application of this approach requires an understanding of the geographic patterns in ecosystems at varying scales of differentiation—the patterns discussed in previous chapters.

Traditionally, USDA Forest Service, Forest Inventory and Analysis (FIA) surveys typically are conducted to estimate and report information according to administrative divisions, such as counties, ownership class, and federal or state forest management districts. This permits stakeholders to rapidly assess priorities and programs under their control or influence. A systematic sample of ground plots over the survey area is conducted to estimate the forest resource. However, surveys involving comprehensive sampling efforts will more accurately characterize unmonitored sites (plots) and discern relationships when samples are stratified according to ecologically similar areas such as ecoregions. O'Brien (1996) and Rudis (1998) found significant differences in the extent and conditions of forest resources among ecoregions in Utah and the southern United States, respectively. To facilitate integration of county-referenced information with areas of similar ecological potential, Rudis (1999) has assigned each county in the conterminous United States to the ecoregion framework.

Global Change.    Considerable attention has been given to the development of a network of stations for monitoring changes in the global environment. Mather and Sdasyuk (1991) suggest two related concepts that should be considered for selecting sites. Obviously, the monitoring site should be representative. Also, stations should be placed where they can detect change. The transitions or ecotones between ecoclimatic regions are potentially suitable for this purpose, with the highest degree of instability of the ecosystems and greatest sensitivity of their components to various forms of pressure occurring there. Since these transition zones are known to respond to changes in climate, it has been postulated that they might be sensitive indicators to climate change (Risser 1993).

In cases in which establishing new monitoring stations is impractical, existing networks and individual studies have to be used. We can compare existing networks to ecosystem maps to see where representation is inadequate and where additional sites are needed. For example, the Long-Term Ecological Research (LTER) sites are located in various ecosystems throughout the United States (Fig. 11.3). By relating the LTER sites to the map, we have a way to establish priorities for new sites. Furthermore, similar ecosystems occur throughout each map unit. By comparing the location of the sites to the map, we can see how far the results of research at a particular site can be extended or transferred to analogous sites within an ecoregion.

**Figure 11.3.** North American ecoregion boundaries and locations of Long-Term Ecological Research (LTER) sites. Compiled by LTER Network Office.

## Design for Sustainability

An understanding of ecosystem geography is relevant to design of sustainable landscapes. It explains how landscapes have evolved and how they may change in the future. Designers and planners use this information to understand landscapes because design and planning involve change. When we design for sustainability, we start by seeing repeated relationships, or patterns, that occur in a particular ecoregion. These patterns reflect formative processes. Ecoregional design is based on the assumption that the processes that shape these patterns can be used to guide planning and design of landscapes, resulting in built environments which are designed differently to best fit each ecoregions unique characteristics. The principles behind pattern-based design are outlined by Barnett and Browning (1995), Van der Ryn and Cowan (1996), Dramstad, Olson, and Forman (1996), Woodward (2000), Bailey (2002), Thayer (2003), and Silbernagel (2005). There are several steps toward implementing this approach.

**Understand Ecosystem Pattern in Terms of Process.**    Again, the identified patterns do not occur randomly, but are linked to the processes that form them. For example, trees that respond to additional moisture are seen repeatedly throughout the arid and semi-arid regions of the western United States. The relationship between pattern and process will vary by region.

**Use Pattern to Design Sustainable Landscapes.**    The natural patterns and processes of a particular region provide essential keys to the sustainability of ecosystems, and can inspire designs for landscapes that sustain themselves. To be sustainable, a designed landscape should imitate the natural ecosystem patterns of the surrounding ecoregion in which they are embedded. By working with nature's design, designers and planners can create landscapes that function sustainably like natural ecosystems. Ecoregional design is the act of understanding the patterns of a region in terms of the processes that shape them and then applying that understanding to design and planning. In addition:

- *Observe how a region functions and try to maintain functional integrity.* Changing the natural patterns by adding subdivisions, road building, or other measures changes ecological functions. In response, animals may change their daily and seasonal movement routes; water flows are changed in their direction and intensity; erosion is accelerated, etc.

- *Maintain diversity by leaving connections and corridors.* Fundamentally, most natural systems are diverse. Therefore, good ecological design will maintain that diversity. Local ecosystems are dependent on the existence of other nearby ecosystems. Therefore, ecosystem diversity depends on leaving some connections and corridors undisturbed.

- *Honor wide-scale ecological processes.* Good ecological design that is sustainable depends on honoring wide-scale ecological processes, including hydrologic cycles, animal movement patterns, and fire regimes.

- *Match development and use of the landscape to its inherent geometry.* By doing so, we can allow existing ecological patterns to work for us. We can use natural drainage patterns to serve as storm drains, wetlands to provide initial treatment of wastewater in place of sewage treatment plants, and indigenous landscaping materials rather than imported ones.

## Spatial Transferability of Models

Another application of this perspective is related to the spatial transferability of models. Ecoregions (macroecosystems, Chapter 2) have two important functions for management.

First, a map of such regions suggests over what area we can apply the knowledge about ecosystem behavior derived from experiments and experience. We can achieve this without too much adjustment, for example, as in silvicultural practices and seed use. Predictive models differ between larger ecosystems; for example, the height-to-age ratio of Norway spruce (*Picea abies*) differs with climate and therefore ecosystems (Fig. 11.4). The climatically defined ecosystem determines which ratio to apply. This is important because if a planner assumes an unsupported growth rate, yield predictions and the forest plans upon which they are based will not reflect reality. The ecoregion map is useful in identifying the geographic extent over which results from site-specific studies (such as growth and yield models) can be reliably extended. Thus the map identifies areas from the spatial transferability of models.

**Figure 11.4.**
Differences in the height-to-age ratio of Norway spruce on similar sites in two different climatic regimes. From Günther (1955) in Barnes (1984), p. 60.

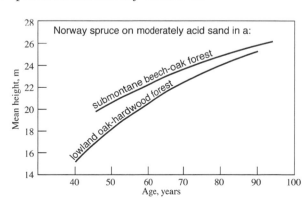

In Canada, studies have found that the height-diameter models of white spruce (*Picea glauca*) were different among different ecoregions (Huang et al. 2000). Incorrectly applying a height-diameter model fitted from one ecoregion to different ecoregions resulted in overestimations between 1 and 29% and underestimations between 2 and 22% of actual growth.

There is another, more compelling example. Each of five regional FIA programs of the eastern United States has developed its own set of volume models, and the models have been calibrated from regions defined by political boundaries corresponding to groups of states rather than ecological boundaries. In some ways, the regional models are quite different. A hypothetical tree shifted a mile in various directions to move from southwest Ohio to southeast Indiana to northern Kentucky could exhibit quite different model-based estimates of volume (Hansen 2002). Growth estimates are likely improved if growth models are calibrated by ecoregions rather than states or FIA regions (Ronald McRoberts, Northern Research Station, personal communication; see also Brooks and Wiant 2007). Other examples of spatial extrapolation in ecology are examined by Miller et al. (2004).

We can apply experience about land use, such as terrain sensitivity to acid rain, suitability for agriculture, and effectiveness of best management practices in protecting fisheries, to similar sites within an ecoregion.

Second, ecoregions identify broad areas in which similar responses may be expected within similarly defined systems. Therefore, we can formulate management policy and apply it on a regional basis rather than on a site-by-site basis. This increases the use of site-specific information and lowers the cost of environmental inventories and monitoring.

A map of ecoregions would have unquestionable value in identifying types of land that will respond in a uniform way to the application of a variety of management practices. A mapping system for identifying landtypes could be useful in current attempts to model the response of wildland areas. Most of these models (e.g., FORPLAN [Johnson et al. 1986] and its replacement, SPECTRUM [Greer and Meneghin 2000]) require that an area undergoing analysis be stratified into homogeneous response units. The ecosystem units described in this book could serve this function.

## Links Between Terrestrial and Aquatic Systems

We cannot regard terrestrial and aquatic components of landscapes as independent systems, because they cannot exist apart from one another. Just as the lower part of a slope exists only in association with the upper, gullies could not form if no watershed existed. The units of a land-scape always comprise connected or associated ecosystems. As stated before, within such systems the diverse ecosystem sites are mutually associated into a whole by the processes of runoff and the migration of chemical elements. Their common history of development also unifies them. Streams are dependent on the terrestrial system in which they are embedded. They therefore have many characteristics in common within a given terrestrial system, including biota and hydrology (Frissell et al. 1986; Swanson et al. 1991).

Aquatic Biota.    A good example of biota that corresponds to a terres-trially defined system is the distribution of the fish northern hog sucker (*Hypentelium nigricans*) (Fig. 11.5). This species is widespread but not uniformly distributed throughout the Mississippi Basin. In Missouri, it is found almost exclusively in the Ozark Uplands landscape, a climatic-landform unit (Pflieger 1971). Benthic invertebrates also coincide nicely with this landscape, expanding its generality to more than fish assem-blages (Rabeni and Doisy 2000).

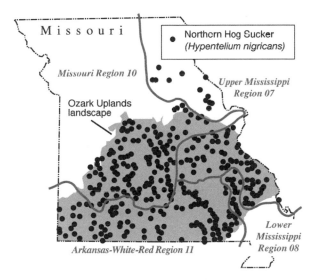

**Figure 11.5.** Distribution of the northern hog sucker in relation to the Ozark Upland landscape and hydrologic units in Missouri. Fish data from Pflieger (1971); hydrologic unit boundaries from U.S. Geological Survey (1979).

A watershed is simply an analytical device based on single criteria (i.e., topographic control of surface waterflow). We need to identify what ecosystem each part of (or all) the watershed is in. That allows us to predict the kinds of streams and associated aquatic organisms that will exist there. Figure 11.5 shows watershed boundaries and one landscape ecosystem. The distribution of the northern hog sucker does not correspond to the watershed boundaries. Comparison, however, identifies areas within the watersheds with similar climatic and landform characteristics and therefore similar aquatic environments. These areas are useful for predicting site conditions and for analysis and management of watersheds.

Other studies have shown that environmental conditions and macroinvertebrate assemblages in boreal headwater streams corresponded relatively well to ecoregion classifications (Heino et al. 2002). Similar results were reported by Harding et al. (1997) in New Zealand. McCreadie and Adler (2006) found that differences in aquatic insect assemblages were clearly detectable across ecoregions in streams from South Carolina, USA. Gallant et al. (2007) used an ecoregion framework to help interpret global amphibian distribution across ecoregion, rather than political, boundaries.

Hydrology.    As discussed in Chapter 4, a strong relationship exists between climatic regime and hydrology. Because ecoregions are based on climate, they should also be hydrologic regions. This approach hypothesizes a relationship between the features of the environment used to delimit the region and the hydrologic properties of the region. One of these properties is hydrologic productivity (i.e., the average normal surface-water runoff). In this case, regions bounded by changes in macrofeatures of the climate are hypothesized to be productively different in important ways. If actual data on hydrologic productivity are assembled for the regions, we can test this hypothesis statistically and evaluate the validity of the regional map. This test is independent of the map, because productivity did not enter into the initial regionalization.

This hypothesis that hydrologic productivity is significantly different from region to region was tested (Bailey 1984). Streamflow data from 53 hydrologic benchmark stations within major ecoregions of the conterminous United States were subjected to discriminant analysis, a technique for analyzing a priori grouped data. These stations are unaffected by urbanization, man-made storage, diversion, or groundwater pumping. The classification results are shown in Table 11.1. The number of stations correctly classified is given as diagonal elements of the matrix, and the numbers of incorrectly classified stations appear as off-diagonal elements. Seven of 53 stations were misclassified. Locations of the stations in relation to domains are shown in Figure 11.6. Eighty-seven percent of grouped stations were correctly classified.

The high percentage of correctly classified stations indicates a high degree of discrimination between the two domains. Thus, the regional ecosystems tested in this study exhibit a high degree of ability to circumscribe a line around a population of stations with similar hydrologic

**Table 11.1.** Classification results for ecoregion domains based on a linear discriminant function using average monthly runoff data[a]

|  |  | Predicted group membership | |
| --- | --- | --- | --- |
| Actual group[b] | No. of stations | 1 | 2 |
| Group 1, Humid temperate | 38 | 32 | 6 |
| Group 2, Dry | 15 | 1 | 14 |

[a]From Bailey (1984).
[b]86.8% of grouped stations was correctly classified.

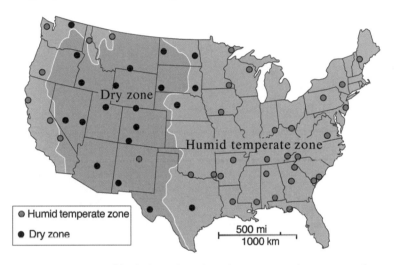

**Figure 11.6.** Location of hydrologic benchmark stations and ecoregion domains of the conterminous United States. From Bailey (1984).

productivity. A group of individuals (or objects) having some common observable characteristics constitutes a population. This provides a better basis to spatially extend data.

The misclassified stations provide a clue to the validity of the map units. For example, most of the misclassified stations are relatively near the dry/humid boundary. We can interpret this to mean that the cores of the regional units are valid but the boundaries, in terms of hydrologic productivity, may need some adjustment. This interpretation would be otherwise if the misclassified stations had been scattered throughout the groups.

The approach taken in this study uses multivariate discriminant analysis to test and validate map units initially recognized and delineated by theoretical considerations. This method differs from previous use of multivariate approaches with land units (Radloff and Betters 1978; Laut and

Paine 1982) that use cluster analysis of grid units to provide the initial map units.

Testing like this also gives a basis for comparing different regionalization schemes (Fig. 11.7).

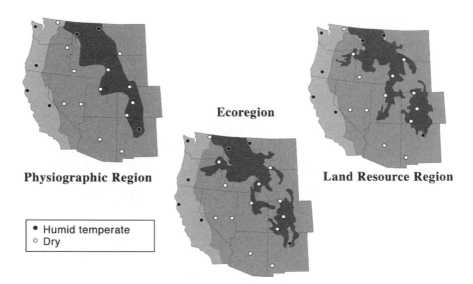

**Figure 11.7.** Distribution of different classes of hydrologic station and their relation to alternative delineations of the Rocky Mountains. Physiographic region from Fenneman (1928); ecoregion from Bailey (1995); land resource region from Soil Conservation Service (Austin 1965).

In New Zealand, Snelder et al. (2005) developed a river environment classification (REC) based on climate and topography in which they expected to discriminate river environments according to differences in water flow regimes. They compared the classification strength of this system to three other classification systems: one based on climate data, a second developed for prediction of low flows, and a third based on an ecoregion classification of New Zealand (Harding and Winterbourn 1997). The ecoregions were defined by overlaying "component" maps of climate, elevation, vegetation, soils, geology and rainfall, which is similar to an approach used by some US analysts (e.g., Omernik 1987). Their study showed that the REC was a stronger classifier of flow regimes than the other classifications they tested. They attribute the increased classification strength of the (REC) to its explicit consideration of the causes of spatial variation in flow regimes among rivers.

## Ecosystem Diversity

Maps of landscape mosaics reveal the relative diversity of ecosystems (Fig. 11.8). Planning and management of diverse and complex landscapes are problematical, whereas more uniform landscapes present relatively simple problems. Solving problems related to land use, such as erosion and revegetation, depends on an understanding of the complexity of the landscape. By knowing the character of the ecosystem mosaic within a landscape and the landscape processes, we can analyze and mitigate the problems associated with management activities.

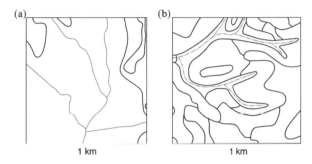

**Figure 11.8.** An example of ecosystem diversity from North Carolina as reflected in soil boundaries: **(a)** is from the Coastal Plain landscape, **(b)** from the Piedmont landscape. From Horton (1967) in Hole (1978).

Rather than occurring randomly, species distributions are sorted in relation to climate and topography (Fig. 11.9). This means that similar climates tend to support similar groups of plants and animals in the absence of human disturbance. Climate influences the distribution of taxa as varied as mammals (Schwartz et al. 2003), spiders (Lightfoot et al. 2008), mosquitoes (Lindsay and Bayoh 2004), and birds (Hanowski et al. 2007). Ecoregional analysis capitalizes on this by identifying climatic and landform factors likely to influence the distribution of species. This analysis uses these factors to define a landscape classification that groups together sites that have similar environmental characteristics (see Fig. 10.21). Such a classification can then be used to indicate sites likely to have similar potential ecosystem character with similar groups of species and similar biological interactions and processes.

One of the major advantages of this approach, as opposed to directly mapping land cover, for example, is its ability to predict the *potential* character of sites where natural ecosystems have been profoundly modified (e.g., by land clearance or fire) or replaced by introduced plants and animals (e.g., pests and weeds).

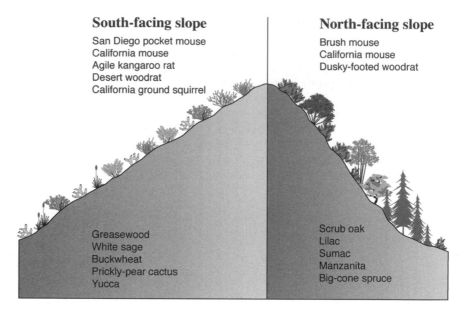

**South-facing slope**
San Diego pocket mouse
California mouse
Agile kangaroo rat
Desert woodrat
California ground squirrel

**North-facing slope**
Brush mouse
California mouse
Dusky-footed woodrat

Greasewood
White sage
Buckwheat
Prickly-pear cactus
Yucca

Scrub oak
Lilac
Sumac
Manzanita
Big-cone spruce

**Figure 11.9.** Mammal and plant communities on south-facing and north-facing slopes in lower San Antonio Canyon, San Gabriel Mountains, California. From Vaughan et al. (2000), data from Vaughn (1954). From Vaughn. *Mammalogy*, 4E. © Brooks/Cole, a part of Cengage Learning, Inc. Reproduced with permission. www.cengage.com/permissions

# Significance to Ecosystem Management

## Management Hierarchies and Ecosystem Hierarchies

Management hierarchies and ecosystem hierarchies correlate so that management strategies, mapping levels, and inventories work well together (Fig. 11.10). This helps form a more consistent and efficient management process.

## Common Permanent Units for Integrated Management

Site maps are general-purpose ecosystem maps. They provide common permanent units for integrated management. We can develop applied ecosystem maps by interpreting and grouping the basic ecosystem units shown on a general-purpose map (Fig. 11.11). For example, we can interpret a general-purpose map to show units with high arboreal productivity and low potential for slope failure. Further interpretation can

| Ecosystem | Management/Analysis |
|---|---|
| Ecoregion | Policy/Strategic (multiforest/forestwide) |
| Landscape Mosaic | Tactical (watershed) |
| Site | Operational (project) |

**Figure 11.10.** Relationship of hierarchies.

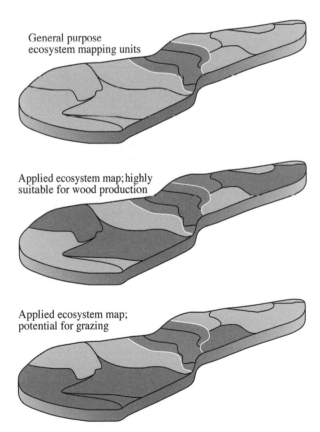

General purpose ecosystem mapping units

Applied ecosystem map; highly suitable for wood production

Applied ecosystem map; potential for grazing

**Figure 11.11.** Common permanent units for integrated management.

place those units into a category of high suitability for wood production. The boundaries remain the same. Applied ecosystem maps will differ only in the interpretation and grouping of the basic ecosystem units.

## Worldwide Application

Each ecoregion is a recurring association that is found in essentially similar form in different parts of the earth. Because they are defined in terms of associations of climate, water, vegetation, and soil, these regions develop a regular pattern over the earth. The middle-latitude combination of continental position, cold winters, warm summers with rainfall during the summer and snow during the winter, for instance, gives the warm continental conditions that may be recognized not only in western Canada and northeastern United States but also in parts of Russia (Fig. 11.12). We can therefore use ecoregion maps to transfer knowledge gained from one part of a continent to another and from one continent to another.

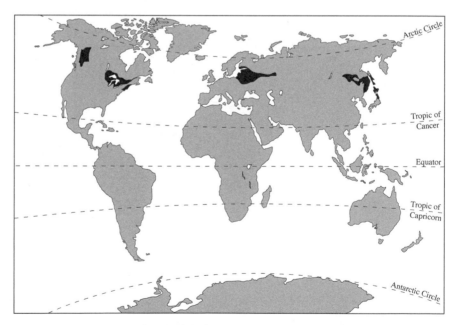

**Figure 11.12.** Generalized global pattern of the warm continental ecoclimatic zone (ecoregion). The map is simplified and drawn on a reduced scale from the author's map (Bailey 1989).

Different areas of the world can have the same or similar ecoregion classifications. For example, ecoregions similar to savanna, steppe, or tundra are found in several continents in addition to North America (Bailey 1998a). Plant species native to a specific ecoregion in North America will likely be well adapted to the same ecoregion on other continents, and vice versa. Thus, ecoregion maps can be used to characterize adaptation of both native and introduced plants at the species level. We can use these map to predict what new harmful organisms (i.e., invasive species) could successfully establish and spread in America if they were to arrive.

However, this approach should be used with caution as ecosystem characteristics have no regional alliance. Because of compensating factors, for example, the same forest type can occur in markedly different ecoregion divisions: ponderosa pine (*Pinus ponderosa*) forest occurs in the northern Rockies and the southwest United States (Fig. 11.13). This distribution does not imply that the climate, topography, soil, and

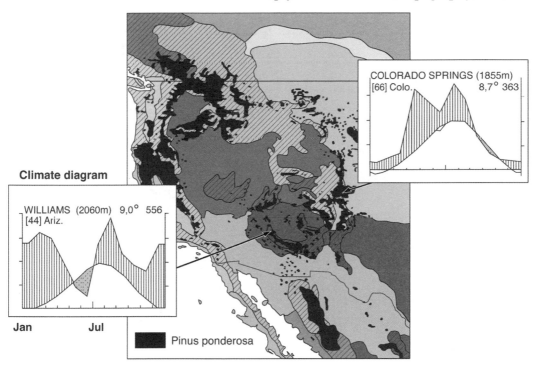

**Figure 11.13.** Ponderosa pine (*Pinus ponderosa*) distribution throughout the various ecoregion divisions (shading) of western North America. Climate diagrams reflect conditions at mountain stations in the tropical/subtropical steppe region (Williams) and in the temperate steppe region (Colorado Springs). Vegetation from U.S. Geological Survey; ecoregions are taken from Bailey (1998b). Climate diagrams redrawn from Walter et al. (1975).

fire regime are the same; the climate diagram for Williams, Arizona, is characterized by late spring drought, whereas Colorado Springs is moist throughout the year.

The distinction between ecoregions is important because of the variable role of ecosystem processes. Some ecoregions have a tendency to large wildfires; the ratio of large to small wildfires decreases from east to west in the conterminous United States (Malamud et al. 2005; Fig. 11.14). Also, fire recurrence interval differs markedly between ecoregions with greater frequency in the West. Pu et al. (2007) analyzed forest fire dataset across North America from satellite data. They found that most fires occurred in the polar eco-domain, subarctic eco-division, and in the tayga (boreal forests), forest-tundra, and open woodlands ecoprovinces in the boreal forests of Canada. The tendency for multiple burns to occur increases with elevation and slope until about 2500 m

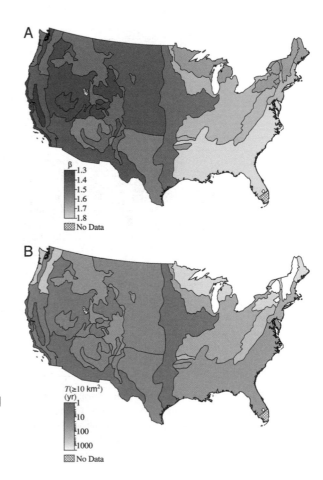

**Figure 11.14.**
Wildfires by ecoregion.
**A**. Ratio of large to small
wildfires; **B**. Fire
recurrence interval.
Source: Malamud et al.
(2005).

elevation and 24° slope, and decreases thereafter. This information suggests that the results of these studies can be used to assess burn probability across the nation to identify areas of high risk, and government agencies could thus better plan responses to wildfire hazards (cf. Littell et al. 2009). Ecoregions could also be used as a baseline from which to assess natural fire regimes, which in turn could be used to abate the threat of fire exclusion and restore fire-adapted ecosystems.

Ecoregion provinces can cover wide zones of latitude. Latitude affects day length during the growing season, length of the growing season, and temperature during both the growing and nongrowing or dormant seasons. Plant populations within an ecosystem often become adapted to their specific latitude via common flowering and maturity characteristics. Populations of a species from different latitudinal zones within an ecoregion can be differentiated by growing the populations in common nurseries located in different latitudes within the ecoregions. These differentiated populations can be referred to as *ecotypes*. A species within an ecoregion is not genetically uniform in regards to adaptation to the entire ecoregion but is stratified into a north-to-south gradient or a high-to-low elevation gradient of ecotypes that are best adapted to their own specific areas of the ecoregions.

The effects of latitude on temperature, winter conditions, and plant growing-season duration within an ecoregion can be modified by geographic features, such as large bodies of fresh water (e.g., the Great Lakes of North America), mountains, and oceans. Plant hardiness zones, such as the USDA Plant Hardiness Zones for North America (Cathey 1990), have been developed to classify plants as to hardiness or survival within ecological zones, which are essentially latitudinal climatic zones modified by nonlatitudinal geographic features. Vogel et al. (2005) described a plant adaptation classification system and an associated map produced by combining ecoregions and hardiness zone maps to develop Plant Adaptation Regions.

Today, various government agencies and international organizations have discovered the value of and the need for ecoregion maps and hence are sponsoring the mapping of large areas, even whole nations. At the same time, environmental problems that are international, interstate, or interprovincial in scope have led to a renewed interest in regionalization. These problems include desertification and long-range transport of air pollutants. Ecoregion maps could form the basis to establish regions of international ecological cooperation for the three American countries (Mexico, Canada, and the United States) that recently negotiated a free trade agreement for the North American region (Szekely 1992). The value of global ecosystem regionalization has been endorsed by the International Union of Forest Research Organizations, and international support and implementation have been recommended (Alston 1987). Bashkin and Bailey (1993) have proposed a new, more accurate and detailed map than the one currently available (Bailey 1989).

# Significance to Research

It is important to link the ecosystem hierarchy with the research hierarchy. In so doing, research structures and ecosystem hierarchies correlate such that research information, mapping levels, and research studies work well together. Comparison of research structures and ecosystem levels can identify gaps in the research network. Table 11.2 shows the relationship between such structures and levels for the US Forest Service research organization.

**Table 11.2.** Relationship between the ecosystem and U.S. Forest Service research hierarchies

| Ecosystem hierarchy* | U.S. Forest Service research organization hierarchy |
| --- | --- |
| Ecoregion | Research Station (multi-ecoregions) |
| Landscape mosaic | Experimental Forest/Range, watershed |
| Site | FIA plot, LTER site, Research Natural Area |

*Source: Bailey 1988a.

At the ecoregional scale, comparison of existing research locations can be compared with ecoregion maps to identify underrepresented regions or gaps in the network. For example, experimental forests or ranges of the USDA Forest Service occur in only 26 of 52 ecoregion provinces (Lugo et al. 2006). Several ecoregions have no research facilities while others have more than one. The greatest number (14) falls within the Laurentian mixed forest ecoregion of the Lake States and Northeast. A more comprehensive analysis could include other types of similar research sites, such as Long Term Ecological Research (LTER) sites, Research Natural Areas, and the like. This analysis could reveal gaps in coverage both across and within ecoregions.

## Restructuring Research Programs

The many useful applications of the study of ecosystem patterns suggest new scientific directions for research and point the way for restructuring research programs. To address critical ecological issues, it is essential to move from the traditional single-scale management and research on plots and stands to mosaics of ecosystems (landscapes and ecoregions) and from streams and lakes to integrated terrestrial–aquatic systems (i.e., geographical ecosystems). FIA thematic maps (e.g., biomass, forest types, etc.) could assist with this.

## Some Research Questions

These research studies reveal useful applications of ecosystem patterns. Many relevant research questions associated with these patterns still remain, including the following: What are the natural ecosystem patterns in a particular ecoregion? What are the effects of climatic variation on ecoregional patterns and boundaries? And what are the relationships between vegetation and landform in different ecoregions? While some have suggested that GIS analysis can assist in answering these questions, that approach should be used with caution because it will help identify pattern, but it cannot generate an understanding of the processes that create these patterns (Bailey 1988b).

Natural Ecosystem Patterns.    Historically, a high level of landscape heterogeneity was caused by natural disturbance and environmental gradients. Now, however, many forest landscapes appear to have been fragmented due to management activities such as timber harvesting, fire suppression, and road construction. To understand the severity of this fragmentation, the nature and causes of the spatial patterns that would have existed in the absence of such activities should be considered. This analysis provides insight into forest conditions that can be attained and perpetuated (Knight and Reiners 2000).

Effects of Climatic Variation.    Climate exerts a very strong effect on ecosystem patterns, and climate change may cause shifts in those patterns (Chapter 8, Neilson 1995). The combination of anthropogenic and cyclical climatic change could yield ecoregions that are much different over time. Therefore, temporal variability is an important research issue. While several researchers are doing work on the effect of climate change on tree species distribution (cf. Iverson and Prasad 2001), others are working on the impact of climatic change on the geography of ecoregions. For example, Jerry Rehfeldt of the Rocky Mountain Research Station (personal communication) has predicted the potential distribution of the American (Mojave-Sonoran) Desert ecoregion under the future climate scenario produced by the IS92a scenario of the Global Climate Model (also know as the general circulation model), with about 5°F warming and 50% increase in precipitation. He has produced maps that show a greatly expanding desert under this scenario: despite the percentage increase in precipitation, the amount of rainfall may fail to keep pace with the increase in temperature so that the climate becomes more arid.

There are limits to the number of monitoring sites that can be established for monitoring changes in the global environment. Obviously, sites should be representative. Stations should also be located where they can detect change. The boundaries between climate-controlled ecoregions are suitable for this purpose. FIA has roughly 160,000 forested sample sites. This criterion could identify a subset of these sites which could be more intensively sampled to provide needed monitoring information.

Relationships Between Vegetation and Landform.    The relationship between vegetation and landform position changes from ecoregion to ecoregion, reflecting the effects of the macroclimate. Species may occupy different positions in the landscape. For the same soil moisture condition but with different topoclimates, tree species change their positions in different regions, for instance (Table 11.3). With these changes, related changes occur in the vigor of other tree species, ecosystem productivity, etc. Knowledge of these differences is important for extending results of research and management experience and for designing sampling networks. These relationships have been extensively studied in some regions (cf. Whittaker and Niering 1965; Peet 1988; Franklin 1998; Odom and McNab 2000) but not in others. Where sufficient studies have been done, these relationships might be modeled and mapped to improve understanding of these ecosystems.

**Table 11.3.** Relationships between vegetation and landform in various ecoregions in Ontario, Canada. From Burger (1976)

| | Topoclimate | | |
| Ecoregion | Hotter | Normal | Colder |
| --- | --- | --- | --- |
| 1 | P | | |
| 2 | P | | P |
| 3 | P | | P |
| 4 | A | P | P |
| 5 | A | | A,P |
| 6 | C | | A,P |
| 7 | | | C,A |

P = *Picea glauca* (white spruce); A = *Acer saccharum* (sugar maple); C = *Carya ovata* (shagbark hickory).

# Conclusion

Ecosystems exist at multiple scales. To address critical ecological issues in ecosystem management, it is essential to move from the traditional single-scale management and research on plots and stands to mosaics of ecosystems and from streams and lakes to integrated terrestrial–aquatic systems (i.e., the multiscale ecosystem). A further challenge is to adopt a common-ecosystem-unit approach that would permit integrated management across functional and administrative lines.

# CHAPTER 12

# Summary and Conclusions

Figure 12.1 summarizes the ideas about the spatial and temporal variability of ecosystems presented in this book. From these ideas, we can draw the following conclusions:

1. We recognize all natural ecosystems by differences in climatic regime. Climate, as a source of energy and moisture, acts as the primary control for the ecosystem. As this component changes, the other components change in response. The primary controls over the climatic effects change with scale. Regional ecosystems are areas of essentially homogeneous macroclimate that biogeographers have traditionally recognized as biomes, life zones, or plant formations.

2. Landform is an important criterion for recognizing smaller divisions within macroecosystems. Landform (with its geologic substrate, its surface shape, and relief) modifies climatic regimes at all scales within macroclimatic zones. It causes the modification of macroclimate to local climate. Thus, landform provides the best means of identifying local ecosystems. At the mesoscale, the landform and landform pattern form a natural ecological unit. At the microscale, we can divide such patterns topographically into slope and aspect units that are relatively consistent in soil moisture regime, soil temperature regime, and plant association (i.e., the homogeneous "site").

3. Present vegetation and land cover are useful for describing the status of the ecosystem in terms of age or disturbance, not to delineate the boundary of the system.

4. Aquatic and riparian systems are closely associated with terrestrial systems and therefore do not need a separate classification, mapping, and/or description mechanism. They are all part of the same landscape ecosystem pattern. All landscapes as ecosystems include both

R.G. Bailey, *Ecosystem Geography*, DOI 10.1007/978-0-387-89516-1_12,
© Springer Science+Business Media, LLC 2009

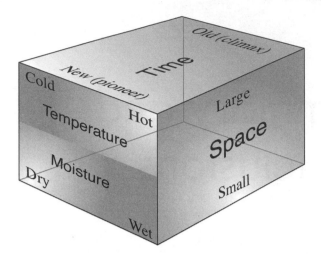

**Figure 12.1.** Spatial and temporal sources of ecosystem variability. Based on a diagram by Cleland et al. (1994).

wet and dry, plus warm and cold, extremes within a region. As such, they are neither terrestrial nor aquatic, but geographic units.

5. The land is conceived as ecosystems, large and small, nested within one another in a hierarchy of spatial sizes. Management objectives and proposed uses determine which sizes are judged important. The aim of useful land classification and mapping is to distinguish appropriately sized ecosystems. Land units will differ significantly from one another, according to resource production capability and the needs of land management.

6. Smaller systems are encompassed in larger systems that control or constrain the operation of the smaller systems. We must examine the relationships between an ecosystem at one scale and ecosystems at smaller or larger scales to predict the effects of management prescriptions on resource outputs. A disturbance to an ecosystem affects smaller component systems.

Therefore, the answer to the question of boundary criteria is that climate, as modified by landform, offers the logical basis for delineating both large and small ecosystems.

# Mapping Criteria

Based on the foregoing analysis, criteria indicative of climatic changes of different magnitude are presented in Table 12.1. Figure 12.2 illus-

**Table 12.1.** Mapping criteria for ecosystem units at a range of scales with examples

| Scale | Name of unit | Criteria | Examples of units Lowland series | Highland series |
|---|---|---|---|---|
| Macro | Ecoregion or zone | Ecoclimatic zone (Köppen 1931) | Temperate semiarid (BSk) | Temperate semiarid regime highlands (H) |
| Meso | Landscape mosaic | Land-surface form (Hammond 1954) | Nearly flat plains (A1) | High mountains (D6) |
| Micro | Site | Topoclimate and soil moisture | Normal topoclimate over moist soil | Normal topoclimate over moist soil |

trates the use of these criteria. Table 12.1 and Figure 12.2 only show the major levels—ecoregion, landscape mosaic, site—not the subdivisions within the levels, such as ecoregion domain, division, and province (see

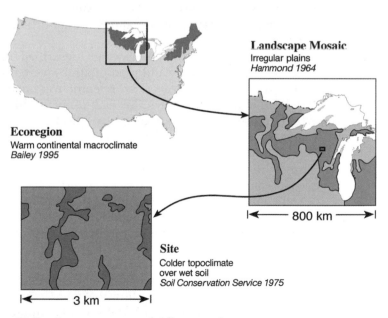

**Figure 12.2.** Ecosystem maps of different scales.

Fig. 6.2). The criteria for delineation are quite different at each of three scales of analysis. They are offered as suggestions to guide the mapping of ecosystems of various sizes. They are not meant to be definitive but to illustrate criteria that appear to be important and that can be used to establish ecosystem boundaries.

With reference to the general principles involved in assigning prime importance at the different scale levels to different criteria (i.e., climate at the macroscale, landform at the mesoscale, and so on), Rowe (1980) has raised the need for a caveat. Although we can map the levels by reference to single physical and biological features, we must always check to ensure that the boundaries have ecological significance. A climatic map showing such key factors as temperature and precipitation is not necessarily an ecological map, unless its boundaries correspond to significant biological boundaries. Likewise, maps of landform, vegetation, and soils are not necessarily ecological maps unless the variation within one map corresponds to the variations with other maps. Before any map is used, it should be thoroughly tested and modified if necessary.

# Boundaries

Ecoregions boundaries are relatively smooth because they are controlled by the macroclimate that lies above the modifying effects of the earth's surface.

Note that in the ecoregion system described in this book, climatic parameters (temperature and precipitation) were used to establish ecoregional differences; however, no attempt was made to use the parameters to establish boundaries. Instead, climatic differences were inferred where discontinuities appeared in physiography (e.g., where plain changes to mountain) and/or vegetation physiognomy (e.g., where tall grass prairie and parkland changes to short-grass steppe or savanna). This is the process that climatologists use to extrapolate their point measurements.

Generally, each climatic region is associated with a single plant formation class (such as broadleaf deciduous forest), characterized by a broad uniformity both in appearance and in composition of the dominant plant species. Of course, not nearly all of the available space is taken up by the plant formation, for the nature of the topography will allow differentiation into many habitats, and the percentage of the climatic region occupied by the formation class that characterizes the formation will depend upon the amount of well-drained upland. Other classes will occupy steep slopes that are hotter and dryer or cooler and moister, as well as bottomlands where the water lies near the surface. A formation class that broadly conforms to the climatic region is termed *zonal*. Local classes correlate with many of the variations from the zonal pattern, and the term *azonal* is applied to these variations. Climatologists and ecosystem geographers ignore these local variations in mapping climatic regions (and, therefore, ecoregions).

In contrast, landscape mosaic boundaries within the ecoregion can be quite irregular because they are controlled by the irregularities of the earth's surface. Usually the control over climatic regime in meso- and micro-ecosystems is strongly physiographic, exerted by the geology and topography. Hence, local ecosystems are best delineated by their basal landforms. Surface differences in shape, substrate, and moisture regime dictate that rain and solar energy will be received and processed in quite different ways by different landforms. Similarly, the much smaller microecosystem units based on topographic facets have their own local climatic regime, indicated by matching of particular soils and biotic communities to slope and aspect.

It should be noted that landscape mosaic boundaries do not affect the placement of ecoregion boundaries but are bounded within the ecoregion. This is an example of how different hierarchical levels constrain one another (Gosz 1993). Landforms, which form the basis for differentiation of meso- and micro-ecosystems, are controlled by internal energy that is independent of macroclimate as a function of latitude and continental position. Landforms are arranged without conforming at all to orderly latitude zones of climate; they cut across them. Because of this, landforms of similar characteristics can be found in various ecoregions, but they will support different ecosystems because of the different climates. In developing the ecoregion concept, a deductive (top–down) approach was used. The upper ecoregion levels of the hierarchy were established, and then physiographic units were sought that correlated with different landform-controlled patterns within the climatically determined ecoregion.

As Rowe (1996) correctly noted, boundaries are recognized by perceived changes in the ecological relationships of vegetation, landform, drainage, and soil, from whose expression climate is inferred. But when vegetation and soil have been drastically disturbed or even destroyed, boundaries between potential ecosystems can still be mapped to coincide with change in those landform characteristics known to regulate the reception and retention of energy and water. At the local scale, for example, the change in contour from convex-upward to concave-downward, from the run-off to the run-in position on hill slopes, is always ecologically significant. In fact, vegetation can often be predicted from site position on slope as well as the shape of the slope (Bolstad et al. 1998; McNab 1989).

# Management Hierarchies and Ecosystem Hierarchies

It is important to link the ecosystem hierarchies with management hierarchies (see Fig. 11.10) because, as pointed out previously, management strategies, mapping levels, and inventories will then work well together. However, three levels of ecological partitioning are not desirable everywhere; there could be two or nine, depending on the kind of question being asked and the scale of the study. However, it is advantageous to have a basic framework consisting of a relatively few units to which all ecological land mappers can relate and within which other units can be defined as required. An example of this is the Forest Service's National Hierarchy of Ecological Units (Fig. 2.3), which has eight levels.

# Human Dimensions

The fundamental knowledge of how ecosystems function and their tolerance for disturbance by human activity should be clearly understood before making ecosystem management decisions. Only then can the tradeoffs necessary to accommodate land uses be weighed in the proper perspective. Social suitability and economic feasibility may then be included to complete the planning process.

# Ecosystem Services

Ecosystem geography is related to biogeography. There are two complimentary *biogeographic* approaches: ecological biogeography and historical biogeography (Crisci et al. 2006). The approach to biogeography presented in this book is based on functional groups of species and environmental constraints, whereas historical biogeography focuses on taxonomic groups and historical biogeographic events. Two locations in the world with similar abiotic characteristics (e.g., precipitation and temperature) may have identical functional groups of organisms and may be considered similar from an ecological point of view, but they may have quite different species composition. For example, climate conditions in the temperate arid and semiarid regions of North and South America are quite similar, and consequently they look similar from an ecological point of view. However, North American prairies and steppes evolved under intensive grazing of bison while large ungulates were absent from South America at that time. One needs to recognize, as Crisci et al. (2006) say,

> Ecological biogeography on its own cannot account for the lack of large ungulates in South America, whereas historical biogeography on its own cannot explain the presence of arid and semiarid vegetation in Central North America. Both subdisciplines are needed in order to achieve a full understanding of biogeography.

The characteristics of ecosystems determine the provision of goods and services to humans. The valuation of such services (such as maintenance of biodiversity) needs to take into consideration changes in biogeography, *both* historical and ecological. Humans modify both environmental constraints (e.g., through climate change) and create new historical events (e.g., through deliberate introduction of species into new areas). These changes may result in new distribution patterns of species or ecosystems and new patterns of the provision of goods and services.

# Ecoregions of the Oceans

Plate 2 (inside back cover) shows regional-scale ecosystem units, or ecoregions, differentiated according to a scheme adapted from Gerhard Schott (1936) as modified by Joerg (1935), James (1936), and Elliott (1954) and using ocean hydrology to indicate the extent of each unit. The units are similar in concept to the continental ecoregions except that the physical characteristics of the water are primary instead of the atmosphere. The map was developed to provide a marine counterpart to the continental ecoregions (Bailey 1989). Understanding the continental systems requires a grasp of the ocean systems that control, through influence on climatic patterns, the continental systems. Two levels or categories of this hierarchy are shown. Of these, the broadest, domains, are based on the major lines of convergence where water masses differing in temperature, salt content, life forms, and color are in contact. Schott recognizes three contrasting types of water, separated by these convergences. Within them, divisions are based largely on the broad natural regions following Dietrich's (1963) system, summarized in Tables A.1 and A.2. Dietrich's system was used with similar results by Hayden et al. (1984) in their proposed regionalization of marine environments.

This classification takes into account the circulation of the oceans, the temperature and salinity, and indirectly, the presence of major zones of upwelling. The characteristic motion of the surface is emphasized because of its influence on temperature. Salinity is generally higher in areas of higher temperature and, therefore, higher evaporation. Marine organisms are usually more abundant in cold waters, which makes the water appear more green.

R.G. Bailey, *Ecosystem Geography*, DOI 10.1007/978-0-387-89516-1,
© Springer Science+Business Media, LLC 2009

**Table A.1.** Natural regions of the oceans[a]

| Dietrich group and types | Ecoregion equivalents |
|---|---|
| B  **Boreal regions** | **Polar domain** (500) |
| Inner boreal (Bi) | Inner polar division    (510) |
| Outer boreal (Bä) | Outer polar division    (520) |
| W  **Westerly drift ocean regions** | **Temperate domain** (600) |
| Poleward of oceanic polar front (Wp) | Poleward westerlies division    (610) |
| Equatorward of polar front (Wä) | Equatorward westerlies division    (620) |
| R  **Horse latitude ocean regions** (R) | Subtropical division    (630) |
| | High salinity subtropical division    (640)[b] |
| F  **Jet current regions** (F) | Jet stream division    (650) |
| M  **Monsoon ocean regions** | |
| Poleward monsoon (Mp) | Poleward monsoon division    (660) |
| | **Tropical domain** (700) |
| Tropical monsoon (Mt) | Tropical monsoon division    (710) |
| | High salinity tropical monsoon division    (720)[b] |
| P  **Trade wind regions** | |
| With poleward current (Pp) | Poleward trades division    (730) |
| With westerly current (Pw) | Trade winds division    (740) |
| With equatorward component (Pä) | Equatorward trades division    (750) |
| A  **Equatorial ocean regions** (Ä) | Equatorial countercurrent division    (760) |

[a]From the Dietrich system (1963).
[b]Not recognized by Dietrich; from Elliott (1954).

More details about the classification system as well as detailed descriptions, illustrations, and examples are presented elsewhere (Bailey 1998a). I describe briefly in the following paragraphs the ecoregion units that appear in the legend to the map.

# Polar Domain

At times or during the winter, ice of the Arctic or Antarctic ocean covers the polar domain. These oceans are characterized, in general, by water that is greenish, low in temperature, low in salt content, and rich in plankton. The duration of ice provides a basis for division into (1) an inner polar zone (division) covered by ice for the entire year and (2) an

**Table A.2.** Definitions and boundaries of the Dietrich system

| | |
|---|---|
| P | (*Passat* in German), persistent westerly setting currents |
| Pä | With 30° equatorward component |
| Pw | Predominantly westerly set |
| Pp | With 30° poleward current |
| Ä | (*Äquator* in German), regions of currents directed at times or all year to the east |
| M | (*Monsun* in German), regions of regular current reversal in spring and autumn |
| Mt | Low-latitude monsoon areas of little temperature variation |
| Mp | Mid- to high (poleward) latitude equivalents of large temperature variations |
| R | (*Ross* in German), at times or all year marked by weak or variable currents |
| F | (*Freistrahlregionen* in German), all-year, geostropically controlled narrow current belts of midlatitude westerly margins of oceans |
| W | (*Westwind* in German), marked by somewhat variable but dominantly east-setting currents all year |
| Wä | Equatorward of oceanic polar front (convergence) |
| Wp | Poleward of oceanic polar front (convergence) |
| B | At times or throughout the year ice covered, in Arctic and Antarctic seas |
| Bi | Entire year covered with ice |
| Bä | Winter and spring covered with ice |

outer polar zone (division) where, with a 50% probability, ice is encountered during winter and spring.

# Temperate Domain

The temperate domain comprises the middle latitudes between the poleward limits of the tropics and the equatorward limits of pack ice in winter. Currents in this region correspond to wind movements around the subtropical atmospheric high-pressure cells. These are the so-called mixed waters of the middle latitudes. The variable direction of the current determines the divisions: (1) a poleward westerlies zone (division) with cold water and sea ice that is poleward of the oceanic polar front; (2) an equatorward westerlies zone (division) that has cool water and is equatorward of the oceanic polar front; (3) a subtropical zone (division) with weak currents of variable directions; (4) a high salinity, subtropical division characterized by excess evaporation over precipitation; (5) a jet steam division characterized by strong, narrow currents that exit during the entire year as the result of discharge from trade wind regions; and (6) a poleward monsoon division of high latitudes with reversal of current (connected with large annual variations in surface temperature).

# Tropical Domain

The tropical domain is characterized by ocean water that is generally blue, high in temperature, high in salt content, and low in organic forms. It is divided into (1) tropical monsoon with regular reversal of the current system (connected with small annual variations in surface temperature); (2) high salinity tropical monsoon with alternating currents; (3) poleward trades with a strong velocity directed toward the poles; (4) trade winds with current moving toward the west; (5) equatorward trades with a strong velocity directed toward the equator, and where currents tend to swing offshore and cold water well up from below; and (6) equatorial countercurrents where currents are directed at times or during the entire year toward the east.

# Shelf

All shallow-shelf areas with depths of 0–200 m are interpreted as shallow variations of the hydrologic zone concerned. A further ecoregional breakdown of coastal and shelf areas of the world is presented by Spalding et al. (2007). A controlling factor approach to classification (see Chapter 3) of estuaries in New Zealand is described by Hume et al. (2007). The classification is based on the principle that particular factors are responsible for environmental processes and patterns that are observed at various spatial scales. The classification differentiates estuaries at four levels of detail. The authors propose that the ocean ecoregions presented in this book, which defines regions of homogeneous climate and oceanic water masses at broad scales, provide an appropriate subdivision of estuaries at Level 1 of the classification, a category based on the ocean ecoregion within which it is located.

# Glossary

*Albedo* reflectivity of the earth environment, generally measured in percentage of incoming radiation.

*Alfisols* soil order consisting of soils with gray to brown surface horizon, medium to high base supply, and subsurface horizons of clay accumulation.

*Algorithm* a sequence of finite mathematical instructions, often used for calculation and data processing.

*Anadromous* going upstream to spawn.

*Anthropogenic* induced or altered by the presence and activities of man.

*Aridisols* soil order consisting of soils with pedogenic horizons, low in organic matter, usually dry.

*Azonal* zonal in a neighboring zone but confined to an extrazonal environment in a given zone.

*Bases* certain cations in the soil that are also plant nutrients; the most important are cations of calcium, magnesium, potassium, and sodium.

*Benthic (zone)* the ecological region at the lowest level of a body of water such as an ocean or a lake, including the sediment surface and some sub-surface layers.

*Biodiversity* variety of life and its processes, including the variety in genes, species, ecosystems, and the ecological processes that connect everything in ecosystems.

*Biota* plant and animal life of a region.

*Biome* largest recognizable subdivision of the terrestrial ecosystem, including total assemblage of plants and animals.

*Block-fault* produced when normal (*near vertical*) *faults* fracture a section of continental crust. Vertical motion of the resulting blocks, sometimes accompanied by tilting, can then lead to high *escarpments*.

These mountains are formed by the earth's crust being stretched and extended by *tensional forces.*

*Boreal forest see* tayga.

*Boroll* suborder of soil order Mollisols; includes Mollisols of cold-winter semiarid plants (steppes) or high mountains.

*Broad-leaved* with leaves other than linear in outline; as opposed to needle-leaved or grasslike (graminoid).

*Brown soil* alkaline soil having a thin, brown surface layer that grades downward into a layer where carbonates have accumulated; developed under grasses and shrubs in semiarid environments.

*Brunizem soil see* prairie soil.

*Calcification* accumulation of calcium carbonate in a soil.

*Carbon sequestration* the process by which carbon sinks (oceans and photosynthesis by plants and animals) remove $CO_2$ from the atmosphere; also known as *$CO_2$ sequestration.*

*Catena see* toposequence.

*Chaparral* sclerophyll scrub and dwarf forest found throughout the coastal mountain ranges and hills of central and southern California.

*Chernozem* fertile, black or dark brown soil under prairie or grassland with lime layer at some depth between 0.6 and 1.5 m.

*Chestnut soil* shortgrass soil in subhumid to semiarid climate with dark brown layer at top, which is thinner and browner than in Chernozem soils, that grades downward to a layer of lime accumulation.

*Chronosequence* changes of a community over time.

*Classification* process of placing objects or phenomena into classes with similar properties.

*Climate* generalized statement of the prevailing weather conditions at a given place, based on statistics of a long period of record.

*Climatic climax* relatively stable vegetation that terminates on zonal soils.

*Climatic regime* seasonality of temperature and moisture.

*Climatic region or zone* group of related climates.

*Climatic type* variety of climate recognized under a system of climate classification.

*Climax* relatively stable state of the vegetation.

*Community* composite of plants, animals, or both that live in association with each other in a given place.

*Compensating factor* factor, or condition, that overrides other factors to bring about the same result.

*Conservation* careful protection, usage, and planned management of living organisms and their vital processes to prevent their depletion, exploitation, destruction, or waste.

*Continentality* tendency of large land areas in midlatitude and high latitudes to impose a large annual temperature range on the air-temperature cycle.

*Cryoboroll* cold Borolls.

*Cryptogram* plant that bears no flowers or seeds but propagates by means of special cells called spores, such as mosses and ferns.

*Cumulative effect* effect on the environment that results from the incremental impact of a proposed action when added to other past, present, and reasonably foreseeable future actions.

*Cyclone* any rotating low-pressure air system.

*Deciduous* woody plants, or pertaining to woody plants, that seasonally lose all their leaves and become temporarily bare-stemmed.

*Desert* supporting vegetation of plants so widely spaced, or sparse, that enough of the substratum shows to give the dominant tone to the landscape.

*Desert soil* shallow gray soils containing little humus; excessive amounts of calcium carbonate at depths less than 0.3 m.

*Division* as defined for use in this book: a subdivision of a domain determined by isolating areas of definite vegetation affinities that fall within the same regional climate.

*Domain* as defined for use in this book: groups of ecoregions with related climates (continents) or water masses (oceans).

*Dry steppe* grassland with 6–8 arid months in each year.

*Ecoclimatic unit* ecosystem based on climate.

*Ecological approach* natural resource planning and management activities that ensure consideration of the relationship among all organisms (including humans) and their environment.

*Ecoregion* as defined for use in this book: a geographic group of landscape mosaics.

*Ecosystem* an area of any size with an association of physical and biological components so organized that a change in any one component will bring about a change in the other components and the operation of the whole system.

*Ecosystem geography* subspecialization of physical geography concerned with the study of the distribution of ecosystems and the processes that have differentiated them.

*Ecosystem management* use of an ecological approach that blends social, physical, economic, and biological needs and values to ensure productive, healthy ecosystems.

*Ecosystem services* benefits from a multitude of resources and processes that are supplied by natural *ecosystems*. Collectively, these benefits are known as ecosystem services and include products like clean drinking water and processes such as the *decomposition* of wastes.

*Ecotone* transition zone between two communities.

*Ecotype* species with wide geographic range that develop locally adapted populations having different limits of tolerance to environmental factors.

*Edaphic* pertaining to soil.

*Edaphic climax* stable community of plants that develops on soils different from those supporting a climatic climax.

*Empirical* dependent on evidence or *consequences* that are *observable* by the *senses*.

*Entisols* soil order consisting of soils without pedogenic horizons.

*Estuary* semi-enclosed *coastal* body of *water* with one or more *rivers* or *streams* flowing into it, and with a free connection to the open *sea*.

*Evapotranspiration* water loss from land by the combined processes of evaporation and transpiration.

*Factor* force that determines a condition in the environment or a direct response from an organism.

*Floodplain* flat surface adjacent to a river channel representing the top of alluvial deposits laid down during floods.

*Flow regime* seasonality of streamflow.

*Forb* broadleaved herb, as distinguished from the grasses.

*Forest* open or closed vegetation with the principal layer consisting of trees averaging more than 5 m in height.

*Forest-tundra* intermingling of tundra and groves or strips of trees.

*Formative process* a set of actions and changes that occur in the landscape as a result of geomorphic, climatic, biotic, and cultural activities.

*Functional* compartmentalization of an organization according to its function (e.g., timber, wildlife, recreation).

*Functional inventory* inventory of natural resources along functional lines (e.g., timber, range, wildlife).

*General circulation model* a class of computer-driven models for *weather forecasting*, understanding *climate* and projecting *climate change*.

*Genetic approach (to mapping ecosystems)* based on the processes that operate to cause the spatial distribution of ecosystems across a range of scales.

*Geographic information system (GIS)* information system for capturing, storing, analyzing, managing and presenting data which are spatially referenced (linked to location).

*Geologic structure* three-dimensional distribution of rock bodies and their folded surfaces.

*Geomorphic* of, or pertaining to the form of the earth's surface.

*Geostrophic* pertaining to deflective force due to rotation of the earth.

*Gleysol or gleyed soil* soil with grayish or bluish fine-textured profile whose color is a consequence of poor aeration that reduces iron compounds.

*Graminoid* related to grasses.

*Gray Brown Podzolic soil* acid soil under broadleaf deciduous forest; has thin organic layer over grayish brown, leached layer; layer of deposition is darker brown.

*Greenhouse gases* gases present in the earth's atmosphere which warm near-surface global temperatures through the *greenhouse effect*.

*Habitat* particular kind of environment in which a species or community lives.

*Habitat type* plant association based on a climax overstory species and indicator understory species.

*Herb* any plant that dies back to the ground surface each year.

*Hierarchy* sequence of sets composed of smaller subsets.

*Histosols* soil order consisting of soils that are organic.

*Inceptisols* soil order consisting of soils with weakly differentiated horizons showing alteration of parent materials.

*Insolation* solar radiation received at earth's surface. It is contracted from incoming solar radiation.

*Imputation* in statistics, imputation is the substitution of some value for a missing data point or a missing component of a data point.

*Integrated classification* following a holistic approach in which both biotic and abiotic components of the ecosystem are considered.

*Intrazonal* exceptional situations within a zone (e.g., on extreme types of soil that override the climatic effect).

*IS92a* one of the emissions scenarios developed in 1992 under the sponsorship of the Intergovernmental Panel on Climate Change. IS92a has been widely adopted as a standard scenario for use in impact assessments.

*Isotherm* line on a map connecting points of equal temperature.

*Kartification* related to the formation of karst (i.e., landscape or topography dominated by surface features of limestone solution and underlain by limestone cavern system).

*k-nearest neighbor algorithm* an imputation technique which estimates unmeasured characteristics on the basis of a number (*k*) of sites which are most similar in respect to the characteristics which are measured.

*Lacrustrine* pertaining to lakes.

*Land* pertaining to the terrestrial part of the earth, as distinguished from sea and air.

*Land capability* level of use an area can tolerate without sustaining permanent damage.

*Landform see* physiography.

*Landscape see* landscape mosaic.

*Landscape mosaic* as defined for use in this book: a geographic group of site-level ecosystems.

*Landscape structure* how the elements of the landscape fit together or are arranged spatially.

*Latisol* reddish, infertile tropical soil in which silica has been leached out, leaving a kaolinitic clay with a high content of iron and aluminum hydroxides.

*Linkages* connections between the components of an ecosystem or the connections between ecosystems.

*Lithic* pertaining to rock.

*Lithology* the physical character of rock.

*Macroclimate* climate that lies just beyond the local modifying irregularities of landform and vegetation.

*Macroinvertebrate* traditionally used to refer to aquatic invertebrates, including insects, crustaceans, molluscs, and worms, which inhabit a river channel, pond, lake, wetland or ocean.

*Mantle* as used by soil scientists: loose, incoherent rock material that rests on the hard or "bed-" rock.

*Mesoclimate* macroclimate modified by local irregularities of landforms, aspect, slope gradient, and elevation.

*Microclimate* climate at or near the ground surface, such as within the vegetation and soil layer.

*Mixed forest* forest with both needle-leaved and broad-leaved trees.

*Model* abstract representation of a system or process. Mathematical models, which use symbolic notation to define relationships describing the system of interest, are commonly used in ecology.

*Mollisols* soil order consisting of soils with a thick, dark-colored, surface-soil horizon, containing substantial amounts of organic matter (humus) and high-base status.

*Monsoon* pertaining to a climatic pattern in which a cool, dry season alternates with a hot, wet season.

*Montane* pertaining to mountain slopes below the alpine belt.

*Moraine* accumulation of rock debris carried and deposited by a glacier.

*Multivariate analysis* methods that concurrently analyze many factors, plus the relationships among those factors.

*Multivariate clustering* statistical procedure for grouping objects by analysis of many factors.

*Muskeg* bogs covered by sphagnum moss.

*Normalized Differences Vegetation Index (NDVI)* simple numerical indicator that can be used to analyze remote sensing measurements and assess whether the target being observed contains live green vegetation or not.

*Oligotrophic* clear-water lake, containing little plankton, often deep and cold and with little thermal stratification, harboring rather poor flora and fauna.

*Open woodland* (also called *steppe forest* and *woodland-savanna*) open forest with lower layers, also open, having the trees or tufts of vegetation discrete but averaging *less* than their diameter apart.

*Orographic* pertaining to mountains.

*Oxisols* soil order consisting of soils that are mixtures principally of kaolin, hydrated oxides, and quartz.

*Pattern-based design* design based on understanding the patterns of a region in terms of process and then applying these patterns to select suitable land-use locations.

*Pedogenic horizon* layer of soil approximately parallel to the land surface and differing from adjacent layers in physical, chemical, and biological properties.

*Pedogenic process* related to soil formation.

*Periglacial process* frost action and frost-produced erosion (e.g., solifluction, frost-heaving, cryoplanation).

*pH* measure of relative acidity or alkalinity, specifically the logarithm of the reciprocal of the hydrogen ion concentration; pH 6.5–7.0 is neutral; higher values are alkaline; lower are acidic.

*Physiognomy (of vegetation)* the outward, superficial appearance of vegetation, without necessary reference to structure or function, even less composition; for example, forest of Douglas fir, of Sitka spruce, of white spruce, and of red fir all have similar physiognomy.

*Physiography* landform (including surface geometry and underlying geologic material).

*Pioneer* an organism getting established on a relatively or absolutely bare area where there is as yet little or no competition.

*Plant association* a kind of plant community represented by stands occurring in places where environments are so closely similar that there is a high degree of floristic uniformity in all layers.

*Plant class* subdivision of the biome based on dominant growth form and cover of the plants that dominate the vegetation.

*Plant subclass* based on morphologic characters, such as evergreen and deciduous habitat, or on adaptation to temperature and water.

*Plant formation* one or more plant communities exhibiting a definite structure and physiognomy; a structural or physiognomic unit of vegetation; for example, a deciduous, broad-leaf forest.

*Plant functional type* groups of species that function similarly but may have quite different species composition.

*Plant hardiness zones* zones developed by the U.S. Department of Agriculture to classify plants to hardiness or survival zones, which are essentially latitudinal climatic zones modified by nonlatitudinal geographic features.

*Plant series* based on individual dominant plant species of the community.

*Plates* pieces of the Earth's *crust* and uppermost mantle, together referred to as the *lithosphere*. The plates are around 100 km (60 miles) thick and consist of two principal types of material: *oceanic crust* (also called *sima,* from *silicon* and *magnesium*) and *continental crust* (*sial*, from silicon and *aluminium*). The composition of the two types of crust differs markedly, with *basaltic* rocks ("mafic") dominating oceanic crust, while continental crust consists principally of lower *density granitic* rocks ("felsic").

*Plate tectonics* the large scale motions of Earth's lithosphere.

*Podzol* soil order consisting of acid soil in which surface soil is strongly leached of bases and clays.

*Potential vegetation* vegetation that would exist if nature were allowed to take its course without human interference.

*Prairie* consists of tallgrasses, mostly exceeding 1 m in height, comprising the dominant herbs, with subdominant forbs (broad-leaved herbs).

*Prairie soil* same as Brunizem; acid grassland soil.

*Primary production* (also called *net primary production*) rate at which carbohydrate is accumulated in the tissues of plants within a given ecosystem; units are grams of dry organic matter per year per square meter of surface area.

*Productivity* rate of dry-matter production by photosynthesis in an ecosystem.

*Province* as defined for use in this book: a subdivision of a division that corresponds to broad vegetation regions, which conform to climatic subzones.

*Radiation* energy transfer in the form of electromagnetic waves.

*Rain shadow (precipitation shadow)* a dry region of land that is *leeward* of a *mountain range* or other geographic feature, with respect to prevailing *wind* direction.

*Relief* difference in elevation between the high and low points of the land surface; it is a function of area.

*Riparian* related to or living on the bank of a river or lake.

*Savanna* closed grass or other predominantly herbaceous vegetation with scattered or widely spaced woody plants, usually including some low trees.

*Scale* level of spatial resolution perceived or considered.

*Sclerophyll or sclerophyllous* refers to plants with predominantly hard stiff leaves that are usually evergreen.

*Semidesert* (also called *half-desert*) area of xerophytic shrubby vegetation with a poorly developed herbaceous lower layer (e.g., sagebrush).

*Sensible heat* heat measurable by a thermometer.

*Shrub* woody plant less than 5 m in height.

*Sierozem* see desert soil.

*Soil* top layer of the earth's surface where rocks have been broken down into relatively small particles through biological, chemical, and physical processes.

*Soil family* defined by physical and chemical characteristics that affect soil use and response to management. Particle size, mineralogy, temperature regime, and depth to root penetration are examples.

*Soil great group* third level of classification of soils, defined by similarities in kind, arrangement, and distinctiveness of horizons, as well as close similarities in moisture and temperature regimes, and base status.

*Soil orders* those ten soil classes forming the highest category in the classification of soils.

*Soil series* basic unit of soil classification, being a subdivision of a family and consisting of soils that are essentially alike in all major characteristics.

*Soil subgroup* defined by characteristics that modify the dominant soil-forming processes.

*Soil suborder* second level of classification of soils, defined by importance of properties that influence soil development and plant growth, such as wetness, parent material, and temperature.

*Spatial* having to do with space or area; place-to-place distribution.

*Spatial hierarchy* geographic grouping of land units based on how they fit together and interact on the land.

*Spodosols* soil order consisting of soils that have accumulations of amorphous materials in subsurface horizons.

*Stand* uninterrupted unit of vegetation, homogeneous in composition and of the same age.

*Steppe* (also called *shortgrass prairie*) open herbaceous vegetation, less than 1 m high, with the tufts or plants discrete yet sufficiently close together to dominate the landscape.

*Substratum* underlayer or stratum, as of earth or rock.

*Subtropical high* one of the semipermanent atmospheric highs of the subtropical high-pressure belt. They are found over the oceans.

*Succession* partial or complete replacement of one plant community with another.

*Sustainable design* the process of prescribing compatible land uses and buildings based on the limits of a place, locally as well as regionally.

*Sustainability* ability of an ecosystem to maintain ecological processes and functions, biological diversity, and productivity over time.

*Systematic sample* sample units that are uniformly distributed over the population.

*Taxa* a group of objects; based on the similarity of properties.

*Tayga (also spelled taiga)* a swampy, parklike savanna with needle-leaved (usually evergreen) low trees or shrubs.

*Terrain* a landsurface, especially with reference to its relief or other natural features.

*Thermokarst* a land surface that forms as *ice*-rich *permafrost* melts. The name is given to very irregular surfaces of marshy hollows and small *hummocks*. These pitted surfaces resemble those formed by solution in some *karst* areas of *limestone*.

*Till plain* an extensive flat *plain* of *glacial till* that forms when a sheet of *ice* becomes detached from the main body of a *glacier* and melts in place, depositing the *sediments* it carried.

*Topoclimate* climate of very small space; influenced by topography.

*Trade winds* two belts of winds that blow almost constantly from easterly directions and are located on the equatorward sides of the subtropical highs.

*Troposphere* lower layer of the atmosphere marked by decreasing temperature, pressure, and moisture with height; the layer in which most day-to-day weather changes occur.

*Tundra* slow-growing low formation, mainly closed vegetation of dwarf shrubs, graminoids, and cryptograms, beyond the subpolar or alpine treeline.

*Tundra soil* cold, poorly drained, thin layers of sandy clay and raw humus; without distinctive soil profiles.

*Utisols* soil order consisting of soils with horizons of clay accumulation and low base supply.

*Vascular plant* fern or seed plant.

*Vegetation* plant covering of an area.

*Vertical zonality* arrangement of climatic zones at different elevations on mountainsides.

*Vertisols* cracking clay soils.

*Watershed* area drained by a river or stream and its tributaries.

*Woodland* cover of trees whose crowns do not mesh, with the result that branches extend to the ground.

*Xerophytic* a plant adapted to an environment characterized by extreme drought.

*Zone* all areas in which the zonal soils have the potential of supporting the same climatic climax plant association.

*Zonal (belt)* a region of the planet that is oriented in an east–west direction with characteristics that are caused by latitudinal variation in solar radiation.

*Zonal soils* well-developed deep soils on moderate surface slope that are well drained.

# Further Reading

Ecosystem geography unites ecology and geography. Lengthy, complete, definitive textbooks exist in both fields, but no books are wholly devoted to the combination. The closest are S. Passarge's *Die Landschaftsgürtel der Erde, Natur und Kultur* (1929), Preston James's *A Geography of Man* (1959), or R. Biasutti's *Il Paesaggio Terrestre* (1962). Although these are noteworthy treatments, they are outdated and out of print. Hidore's *Physical Geography: Earth Systems* (1974) and Strahler and Strahler's *Elements of Physical Geography* (1976) identify environmental regions on a global scale but do not go beyond that to identify geographic patterns of ecosystems within regions. Billing's *Plants and the Ecosystem* (1964), Bennett's *Man and Earth's Ecosystems* (1975), and Walter's *Vegetation of the Earth and Ecological Systems of the Geobiosphere* (1984) are similarly focused.

The only general text on landscape geography is A. G. Isachenko's *Principles of Landscape Science and Physical-Geographic Regionalization* (1973). Landscape science involves the delineation, description, and analysis of relatively homogeneous units of land at the local or regional scale. As it deals with evolution and process of the total landscape as well as of its components, it is similar in concept to ecosystem geography.

Several works deal with the effect of climate on ecosystem differentiation at multiple scales. Examples of some of the most fundamental contributions in this area are Pierre Dansereau's *Biogeography—An Ecological Perspective* (1957), Peter Haggett's *Geography: A Modern Synthesis* (1972), and Angus Hills's "Comparison of Forest Ecosystems (Vegetation and Soil) in Different Climatic Zones" (1960b). A recent treatment of this topic is Gordon Bonan's advanced text, *Ecological Climatology* (2002). Landform influences on ecosystem patterns are presented in Hunt's *Geology of Soils: Their Evolution, Classification, and Uses* (1972) and Kruckeberg's *Geology and Plant Life* (2002).

Schemes for recognizing ecosystems at multiple scales are summarized in Miller's "The Factor of Scale: Ecosystem, Landscape Mosaic, and Region" (1978), Rowe and Sheard's "Ecological Land Classification: A Survey Approach" (1981), and Godron's "The Natural Hierarchy of Ecological Systems" (1994).

Ideas related to those discussed in this book have appeared in several texts. An example is Richard Huggett's excellent *Geoecology* (1995), which deals with how animals, plants, and soils interact with one another and with the terrestrial spheres (atmosphere, hydrosphere, troposphere, and lithosphere), creating landscape systems or 'geoecosystems'. For information on the geoecosystems studied by landscape ecologists, a good starting point is the textbook by Forman and Godron (1986), *Landscape Ecology*. H. Leser's *Landschaftsökologie* (1976) gives a comprehensive treatment in German of the European approach to landscape ecology. J. Schultz's *The Ecozones of the World: The Ecological Divisions of the Geosphere* (1995), now in its second edition, provides a detailed look at the world's ecozones. The book by Monica Turner and her colleagues (2001) emphasizes the interaction between spatial pattern and ecological process across a range of scales. An overview of ecology from a scale-linking perspective is provided by T. Allen and T. Hoekstra in their *Toward a Unified Ecology* (1992). My second book, *Ecoregions: The Ecosystem Geography of the Oceans and Contients* (Bailey 1998a), applies the principles described in this book to delineate and characterize the major terrestrial and aquatic zones of the Earth. *Ecoregion-Based Design for Sustainability* (Bailey 2002), the third in the series, goes on to explain the utility of the ecoregion concept in the management and design of sustainable landscapes.

# Bibliography

Akin, W.E. 1991. Global patterns: climate, vegetation, and soils. Norman, OK: University of Oklahoma Press. 370p.

Albert, D.A.; Denton, S.R.; Barnes, B.V. 1986. Regional landscape ecosystems of Michigan. Ann Arbor, MI: University of Michigan. 32p. with separate map at 1:1,000,000.

Allen, T.F.H.; Starr, T.B. 1982. Hierarchy: perspective for ecological complexity. Chicago: University of Chicago Press. 310p.

Allen, T.F.H.; Hoekstra, T.W. 1992. Toward a unified ecology. New York: Columbia University Press. 384p.

Alston, R.M. 1987. Inventorying and monitoring endangered forests: a critical report on the 1985 IUFRO Conference. Journal of Environmental Management. 25: 181–189.

Anderson, J.R.; Hardy, E.E.; Roach, J.T.; Witmer, R.E. 1976. A land use and land cover classification system for use with remote sensor data. Prof. Paper 964. Washington, DC: U.S. Geological Survey. 28p.

Atwood, W.W. 1940. The physiographic provinces of North America. Boston: Ginn. 536p.

Austin, M.E. 1965. Land resource regions and major land resource areas of the United States (exclusive of Alaska and Hawaii). Agric. Handbook 296. Washington, DC: USDA Soil Conservation Service. 82p. with separate map at 1:7,500,000.

Bailey, R.G. 1976. Ecoregions of the United States. Ogden, UT: USDA Forest Service. Intermountain Region. 1:7,500,000; colored.

Bailey, R.G. 1981. Integrated approaches to classifying land as ecosystems. In: P. Laban (ed.). Proceedings of the workshop on land evaluation for forestry. November 10–14, 1980. Wageningen, The Netherlands. Wageningen: International Institute for Land Reclamation and Improvement. pp. 95–109.

Bailey, R.G. 1983. Delineation of ecosystem regions. Environmental Management. 7: 365–373.

Bailey, R.G. 1984. Testing an ecosystem regionalization. Journal of Environmental Management. 19: 239–248.

Bailey, R.G. 1985. The factor of scale in ecosystem mapping. Environmental Management. 9: 271–276.

Bailey, R.G. 1987. Suggested hierarchy of criteria for multi-scale ecosystem mapping. Landscape and Urban Planning. 14: 313–319.

Bailey, R.G. 1988a. Ecogeographic analysis: a guide to the ecological division of land for resource management. Misc. Publ. No. 1465. Washington, DC: USDA Forest Service. 16p.

Bailey, R.G. 1988b. Problems with using overlay mapping for planning and their implications for geographic information systems. Environmental Management. 12: 11–17.

Bailey, R.G. 1989. Explanatory supplement to ecoregions map of the continents. Environmental Conservation. 16: 307–309, with separate map at 1:30,000,000.

Bailey, R.G. 1991. Design of ecological networks for monitoring global change. Environmental Conservation. 18: 173–175.

Bailey, R.G. 1994. Map. Ecoregions of the United States (rev.). Washington, DC: USDA Forest Service. Scale 1:7,500,000.

Bailey, R.G. 1995. Description of the ecoregions of the United States. 2nd ed. rev. and expanded (1st ed. 1980). Misc. Publ. No. 1391 (rev.). Washington, DC: USDA Forest Service. 108p. with separate map at 1:7,500,000.

Bailey, R.G. 1997. Map. Ecoregions of North America (rev.). Washington, DC: USDA Forest Service. Scale 1:15,000,000.

Bailey, R.G. 1998a. Ecoregions: the ecosystem geography of the oceans and continents. New York: Springer-Verlag. 176p.

Bailey, R.G. 1998b. Ecoregions map of North America: explanatory note. Misc. Publ. 1548. Washington, DC: USDA Forest Service. 10p.

Bailey, R.G. 2002. Ecoregion-based design for sustainability. New York: Springer. 222p.

Bailey, R.G. 2005. Identifying ecoregion boundaries. Environmental Management. 34(Suppl.1): S14–S26.

Bailey, R.G.; Cushwa, C.T. 1981. Ecoregions of North America. FWS/OBS-81/29. Washington, DC: U.S. Fish and Wildlife Service. 1:12,000,000; colored.

Bailey, R.G.; Pfister, R.D.; Henderson, J.A. 1978. Nature of land and resource classification: a review. Journal of Forestry. 76: 650–655.

Bailey, R.G.; Avers, P.E.; King, T.; McNab, W.H. (eds.). 1994. Ecoregions and subregions of the United States. Washington, DC: USDA Forest Service. 1:7,500,000; colored with supplementary table of map unit descriptions, compiled and edited by W.H. McNab and R.G. Bailey.

Bailey, R.W. 1941. Climate and settlement of the arid region. In: 1941 Yearbook of agriculture. Washington, DC: U.S. Department of Agriculture. pp. 188–196.

Barnes, B.V. 1984. Forest ecosystem classification and mapping in Baden-Württemberg, West Germany. In: J.G. Bockheim (ed.). Proceedings, forest land classification: experiences, problems, perspectives. March 18–20, 1984. Madison, Wisconsin. Madison, WI: Department of Soil Science, University of Wisconsin. pp. 49–65.

Barnett, D.L.; Browning, W.D. [illustrations by J.L. Uncapher] 1995. A primer on sustainable building. Snowmass, CO: Rocky Mountain Institute. 135p.

Barnes, B.V.; Pregitzer, K.S.; Spies, T.A.; Spooner, V.H. 1982. Ecological forest site classification. Journal of Forestry. 80: 493–498.

Barry, R.G. 1992. Mountain weather and climate. 2nd ed. London: Routledge. 402p.

Bashkin, V.N.; Bailey, R.G. 1993. Revision of map of ecoregions of the world (1992–95). Environmental Conservation. 20: 75–76.

Bazilevich, N.I.; Rodin, L.Y.; Rozov, N.N. 1971. Geographical aspects of biological productivity. Soviet Geography. 12: 293–317.

Bear, F.E.; Pritchard, W.; Akin, W.E. 1986. Earth: the stuff of life. 2nd ed. Norman, OK: University of Oklahoma Press. 318p.

Beckinsale, R.P. 1971. River regimes. In: R.J. Chorley (ed.). Introduction to physical hydrology. London: Methuen. pp. 176–192.

Bennett, C.F. 1975. Man and earth's ecosystems. New York: John Wiley. 331p.

Berg, L.S. 1947. Geograficheskiye zony Sovetskogo Soyuza (geographical zones of the Soviet Union). vol. 1, 3rd ed. Moscow: Geografgiz.

Berghaus, H. 1845. Physikalischer Atlas. Gotha, Germany: Justus Perthes Verlag. vol. 1.

Bernert, J.A.; Eilers, J.M.; Sullivan, T.J.; Freemark, K.E.; Ribic, C. (1997) A quantitative method for delineating regions: an example from the Western Corn Belts Ecoregions of the USA. Environmental Management. 21: 405–420.

Blasi, C.; Carranza, M.L.; Frondoni, R.; Rosati, L. 2000. Ecosystem classification and mapping: a proposal for Italian landscapes. Applied Vegetation Science. 3: 233–242.

Biasutti, R. 1962. Il paesaggio terrestre. 2d ed. Torino: Unione Tipografico. 586p.

Billings, W.D. 1964. Plants and the ecosystem. Belmont, CA: Wadsworth. 154p.

Bockheim, J.G. 2005. Soil endemism and its relation to soil formation theory. Geoderma 129: 109–124.

Bogorov, V.G. 1962. Problems of the zonality of the world ocean. In: C.D. Harris (ed.). Soviet geography, accomplishments and tasks. Occasional Publ. No. 1. New York: American Geographical Society. pp. 188–194.

Bolstad, P.V.; Swank, W.; Vose, J. 1998. Predicting Southern Appalachian overstory vegetation with digital terrain data. Landscape Ecology. 13: 271–283.

Bonan, G.B. 2002. Ecological climatology. Cambridge, UK: Cambridge University Press. 678p.

Borchert, J.F. 1950. The climate of the central North American grassland. Annals Association of American Geographers. 40: 1–39.

Bourne, R. 1931. Regional survey and its relation to stocktaking of the agricultural and forest resources of the British Empire. Oxford Forestry Memoirs 13. Oxford: Clarendon Press. 169p.

Bowman, I. 1911. Forest physiography, physiography of the U.S. and principal soils in relation to forestry. New York: John Wiley. 759p.

Box, E.O. 1981. Macroclimate and plant forms: an introduction to predictive modeling in phytogeography. The Hague: Dr. W. Junk Publishers. 258p.

Branson, F.A.; Shown, L.M. 1990. Contrasts of vegetation, soils, microclimates, and geomorphic processes between north- and south-facing slopes on Green

Mountain near Denver, Colorado. U.S. Geological Survey Water-Resources Investigations Report 89-4094.

Breymeyer, A.I. 1981. Monitoring of the functioning of ecosystems. Environmental Monitoring and Assessment. 1: 175–183.

Brooks, J.R.; Wiant, H.V. 2007. Ecoregion-based local volume equations for Appalachian hardwoods. Northern Journal of Applied Forestry. 25(2): 87–92.

Brown, J.H. 1995. Macroecology. Chicago: University of Chicago Press. 269p.

Budyko, M.I. 1974. Climate and life (English edition by D.H. Miller). New York: Academic Press. 508p.

Bunnett, R.B. 1968. Physical geography in diagrams. New York: Frederick A. Praeger. 172p.

Burger, D. 1976. The concept of ecosystem region in forest site classification. In: Proceedings, International Union of Forest Research Organizations (IUFRO), XVI world congress, division I; 20 June–2 July 1976; Oslo, Norway. Oslo, Norway: IUFRO. pp. 213–218.

Cathey, H.M. 1990. USDA plant hardiness zone map. Washington, DC: U.S. Department of Agriculture. USDA Misc. Pub. No. 1475.

Chen, J.; Saunders, S.C.; Crow, T.R.; Naiman, R.J.; et al. 1999. Microclimate in forest ecosystem and landscape ecology. Bioscience. 49: 288–297.

Christian, C.S.; Stewart, G.A. 1968. Methodology of integrated surveys. In: Aerial surveys and integrated studies. Proceedings Toulouse Conference 1964. Paris: UNESCO. pp. 233–280.

Christopherson, R.W. 2000. Geosystems: an introduction to physical geography. 4th ed. Upper Saddle River, NJ: Prentice Hall. 626p.

Claussen, M.; Esch, M. 1994. Biomes computed from simulated climatologies. Climate Dynamics. 9: 235–243.

Cleland, D.T.; Crow, T.R.; Hart, J.B.; Padley, E.A. 1994. Resource management perspective: remote sensing and GIS support for defining, mapping and managing ecosystems. In: V. Alaric Sample (ed.). Remote sensing and GIS in ecosystem management. Washington, DC: Island Press. pp. 218–242.

Cleland, D.T.; Freeouf, J.A.; Keys, Jr., J.E.; Nowacki, G.J.; et al. 2005. Map. Ecological subregions: sections and subsections of the conterminous United States. CD-ROM. Washington, DC: USDA Forest Service. Scale 1:3,500,000. Available: http://svinetfc4.fs.fed.us/research/section/index.html

Corbel, J. 1964. L'erosion terrestre etude quantitative (methods-technques-resultats). Annales de Geographie 73: 385–412.

Cowardin, L.M.; Carter, V.; Golet, F.C.; LaRoe, E.T. 1979. Classification of wetlands and deep-water habitats of the United States. FWS/OBS-79/31. Washington, DC: U.S. Fish and Wildlife Service. 103p.

Crisci, J.V.; Sala, O.E.; Katinas, L.; Posadas, P. 2006. Bridging historical and ecological approaches in biogeography. Australian Systematic Botany. 19: 1–10.

Crowley, J.M. 1967. Biogeography. Canadian Geographer. 11: 312–326.

Damman, A.W.H. 1979. The role of vegetation analysis in land classification. Forestry Chronicle. 55: 175–182.

Dansereau, P. 1957. Biogeography—an ecological perspective. New York: Ronald Press. 394p.

Dasmann, R.F. 1972. Towards a system for classifying natural regions of the world and their representation by national parks and reserves. Biological Conservation. 4: 247–255.

Daubenmire, R. 1943. Vegetation zonation in the Rocky Mountains. Botanical Review. 9: 325–393.

Daubenmire, R. 1968. Plant communities: a text book on plant synecology. New York: Harper & Row. 300p.

Davis, L.S. 1980. Strategy for building a location-specific, multipurpose information system for wildland management. Journal of Forestry. 78: 402–408.

Davis, W.M. 1899. The geographical cycle. Geographical Journal. 14: 481–504.

De Castro, M.; Gallardo, C.; Jylha, K.; Tuomenvirta, H. 2007. The use of a climate-type classification for assessing climate change effects in Europe from an ensemble of nine regional climate models. Climatic Change. 81: 329–341.

de Laubenfels, D.J. 1970. A geography of plants and animals. Dubuque, IA: Wm. C. Brown. 133p.

Delvaux, J.; Galoux, A. 1962. Les territoires écologiques du sud-est belge. Centre d'Ecologie generale. Travaux hors serie. 311p.

Diaz, H.F.; Eischeid, J.K. 2007. Disappearing "alpine tundra" Köppen climatic type in the western United States. Geophysical Research Letters. 34: L18707.

Dietrich, G. 1963. General oceanography: an introduction. New York: John Wiley. 588p.

Dix, R.L.; Smeins, F.E. 1967. The prairie, meadow, and marsh vegetation of Nelson County, North Dakota. Canadian Journal of Botany. 45: 21–58.

Dodd, M.B.; Lauenroth, W.K.; Burke, I.C.; Chapman, P.L. 2002. Associations between vegetation patterns and soil texture in the shortgrass steppe. Plant Ecology. 158: 127–137.

Dokuchaev, V.V. 1899. On the theory of natural zones. Sochineniya (collected works). vol. 6, Moscow-Leningrad: Academy of Sciences of the USSR, 1951.

Dolan, B.J.; Parker, G.R. 2005. Ecosystem classification in a flat, highly fragmented region of Indiana, USA. Forest Ecology and Management. 219: 109–131.

Dramstad, W.E.; Olson, J.D.; Forman, R.T.T. 1996. Landscape ecology principles in landscape architecture and land-use planning. Washington, DC: Island Press. 80p.

Driscoll, R.S.; Merkel, D.L.; Radloff, D.L.; Snyder, D.E.; Hagihara, J.S. 1984. An ecological land classification framework for the United States. Misc. Publ. 1439. Washington, DC: U.S. Department of Agriculture. 56p.

Dryer, C.R. 1919. Genetic geography: the development of the geographic sense and concept. Annals Association of American Geographers. 10: 3–16.

ECOMAP. 1993. National hierarchical framework of ecological units. Washington, DC: USDA Forest Service. 20p.

Ecoregions Working Group. 1989. Ecoclimatic regions of Canada, first approximation. Ecological Land Classif. Series No. 23. Ottawa: Environment Canada. 119p. with separate map at 1:7,500,000.

Elliott, F.E. 1954. The geographic study of the oceans. In: P.E. James and C.F. Jones (eds.). American geography: inventory & prospect. Syracuse, NY: Association of American Geographers by Syracuse University Press. pp. 410–426.

Emanuel, W.R.; Shugart, H.H.; Stevenson, M.P. 1985. Climatic change and the broad-scale distribution of terrestrial ecosystem complexes. Climate Change. 7: 29–43.

Fairbridge, R.W. 1963. Africa ice-age aridity. In: Nairn, A.E.M. (ed.). Problems in paleoclimatology. London: Wiley. pp. 356–363.

FAO. 1984. Land evaluation for forestry. FAO Forestry Paper 48. Rome, Italy: Food and Agriculture Organization of the United Nations (FAO). 123p.

FAO. 2001. Global forest resources assessment 2000. FAO Forestry Paper 140. Rome, Italy: Food and Agriculture Organization of the United Nations (FAO). 479p.

FAO/UNESCO. 1971–1978. FAO/UNESCO soil map of the world 1:5 million. North America, South America, Mexico and Central America, Europe, Africa, South Asia, North and Central Asia, Australia. Paris: UNESCO.

Fenneman, N.M. 1928. Physiographic divisions of the United States. Annals Association of American Geographers. 18: 261–353.

Ferguson, B.K. 1992. Landscape hydrology, a component of landscape ecology. Journal of Environmental Systems. 21: 193–205.

Forman, R.T.T.; Godron, M. 1986. Landscape ecology. New York: John Wiley. 619p.

Fosberg, F.R.; Garnier, B.J.; Küchler, A.W. 1961. Delimitation of the humid tropics. Geographical Review. 51: 333–347, with separate map at 1:60,000,000.

Franklin, J. 1998. Predicting the distribution of shrub species in California chaparral and coastal sage communities from climate and terrain-derived variables. Journal Vegetation Science. 9: 733–748.

Frissell, C.A.; Liss, W.J.; Warren, C.E.; Hurley, M.C. 1986. A hierarchical framework for stream habitat classification: viewing streams in a watershed context. Environmental Management. 10: 199–214.

Funk, J.L. 1970. Warm-water streams. In: N.G. Benson (ed.). A century of fisheries in North America. Washington, DC: American Fisheries Society. pp. 141–152.

Gallant, A.L.; Binnian, E.F.; Omernik, J.M.; Shasby, M.B. 1995. Ecoregions of Alaska. US Geological Survey Professional Paper 1567. Washington, DC: US Geological Survey, with separate map at 1:5,000,000.

Gallant, A.L.; Klaver, R.W.; Casper, G.S.; Lannoo, M.J. 2007. Global rates of habitat loss and implications for amphibian conservation. Copeia. 4: 967–979.

Gaussen, H. 1954. Théorie et classification des climats et microclimats. 8me Congr. Internat. Bot. Paris, Sect. 7 et 3. pp. 125–130.

Geiger, R. 1965. The climate near the ground. (trans.). Cambridge, MA: Harvard University Press. 611p.

Gerasimov, I.P. (ed.). 1964. Types of natural landscapes of the earth's land areas. Plate 75. In: Fiziko-geograficheskii atlas mira (physico-geographic atlas of the world). Moscow: USSR Acad. Sci. and Main Administration of Geodesy and Cartography. Scale 1:80,000,000.

Gersmehl, P.J. 1980. Productivity ratings based on soil series: a methodology critique. Professional Geographer. 32: 158–163.

Gersmehl, P.J. 1981. Maps in landscape interpretation. Cartographica. 18: 79–115.

Gersmehl, P.; Napton, D.; Luther, J. 1982. The spatial transferability of resource interpretations. In: T.B. Braun (ed.). Proceedings, national in-place resource inventories workshop, August 9–14, 1981. University of Maine, Orono. Washington, DC: Society of American Foresters. pp. 402–405.

Godron, M. 1994. The natural hierarchy of ecological systems. In: F. Klijn (ed.). Ecosystem classification for environmental management. Dordrecht: Kluwer Academic Publishers. pp. 69–83.

Goff, F.G.; Baxter, F.P.; Shugart, Jr., H.H. 1971. Spatial hierarchy for ecological modeling. EDFB Memo Rep. 71–41. Oak Ridge, TN: Oak Ridge National Laboratory. 12p.

Gosz, J.R. 1993. Ecotone hierarchies. Ecological Applications. 3: 369–376.

Gosz, J.R.; Sharpe, P.J.H. 1989. Broad-scale concepts for interactions of climate, topography, and biota at biome transitions. Landscape Ecology. 3: 229–243.

Greer, K.; Meneghin, B. 2000. Spectrum: an analytical tool for building natural resource management models. In: J.M. Vasievich, J.S. Fried, and L.A. Leefers (eds.). Seventh symposium on systems analysis in forest resources; 1997 May 28–31; Traverse City, MI. Gen. Tech. Rep. NC-205. St. Paul, MN: USDA Forest Service, North Central Research Station. 470p.

Gregg, R.E. 1964. Distribution of the ant genus Formica in the mountains of Colorado. In: H.G. Rodeck (ed.). Natural history of the Boulder area. Leaflet, University of Colorado Museum 13: 59–69.

Grigor'yev, A.A. 1961. The heat and moisture regime and geographic zonality. Soviet Geography: Review and Translation. 2: 3–16.

Günther, M. 1955. Untersuchungen über das Ertragsvermögen der Hauptholzarten im Bereich verschiedener des württembergischen Neckarlandes. Mitt. Vereins f. forstl. Standortsk. u. Forstpflz. 4: 5–31.

Hack, J.T.; Goodlet, J.C. 1960. Geomorphology and forest ecology of a mountain region in the central Appalachians. Prof. Paper 347. Washington, DC: U.S. Geological Survey. 66p.

Haggett, P. 1972. Geography: a modern synthesis. New York: Harper & Row. 483p.

Halpin, P.N. 1994. Latitudinal variation in the potential response of mountain ecosystems to climatic change In: M. Beniston (ed.). Mountain environments in changing climates. London & New York: Routledge. pp. 180–203.

Hammond, E.H. 1954. Small-scale continental landform maps. Annals Association of American Geographers. 44: 33–42.

Hammond, E.H. 1964. Classes of land-surface form in the forty eight states, USA. Annals Association of American Geographers. 54. Map supplement no. 4, scale 1:5 million.

Hanowski, J.; Danz, N.; Howe, R.; Niemi, G.; Regal, R. 2007. Consideration of geography and wetland geomorphic type in the development of Great Lakes coastal wetland bird indicators. EcoHealth. 4: 194–205.

Hansen, M. 2002. Volume and biomass estimation in FIA: national consistency vs. regional accuracy In: R.E. McRoberts, G.A. Reams, P.C. Van Deusen, and J.W. Moser (eds.). Proceedings of the third annual forest inventory and analysis symposium; Gen. Tech. Rep. NC-230. St. Paul, MN: USDA Forest Service, North Central Research Station. pp. 109–120.

Harding, J.S.; Winterbourn, M.J. 1997. An ecoregion classification of the South Island, New Zealand. Journal of Environmental Management. 51: 275–287.

Harding, J.S.; Winterbourn, M.J.; McDiffett, W.F. 1997. Stream fauna and ecoregions in South Island, New Zealand: do they correspond? Archiv für Hydrobiologie. 140(3): 289–307.

Hare, F.K. 1950. Climate and zonal divisions of the boreal forest formation in eastern Canada. Geographical Review. 40: 615–635.

Hargrove, W.W.; Luxmoore, R.J. 1998. A clustering technique for the generation of customizable ecoregions. Proceedings ESRI Arc/INFO users conference. Available at http://research.esd.ornl.gov/~hnw/esri98/

Hayden, B.P.; Ray, G.C.; Dolan, R. 1984. Classification of coastal and marine environments. Environmental Conservation. 11: 199–207.

Heino, J.; Muotka, T.; Paavola, R.; Hämäläinen, H.; Koskenniemi, E. 2002. Correspondence between regional delineations and spatial patterns in macroinvertebrate assemblages of boreal headwater streams. Journal North American Benthological Society. 21: 397–413.

Hembree, C.H.; Rainwater, F.H. 1961. Chemical degradation on opposite flanks of the Wind River Range, Wyoming. Water-Supply Paper 1535-E. Washington, DC: U.S. Geological Survey. 9p.

Herbertson, A.J. 1905. The major natural regions: an essay in systematic geography. Geography Journal. 25: 300–312.

Hidore, J.J. 1974. Physical geography: earth systems. Glenview, IL: Scott, Foresman and Co. 418p.

Hidore, J.J.; Oliver, J.E. 1993. Climatology: an atmospheric science. New York: Macmillan. 423p.

Hills, A. 1952. The classification and evaluation of site for forestry. Res. Rep. 24. Toronto: Ontario Department of Lands and Forest. 41p.

Hills, G.A. 1960a. Regional site research. Forestry Chronicle. 36: 401–423.

Hills, G.A. 1960b. Comparison of forest ecosystems (vegetation and soil) in different climatic zones. Silva Fennica. 105: 33–39.

Hills, G.A. 1976. An integrated interactive holistic approach to ecosystem classification. In: J. Thie and G. Ironside (eds.). Ecological (biophysical) land classification in Canada. Ecological Land Classification Series No. 1. Ottawa: Environment Canada. pp. 73–97.

Holdridge, L.R. 1947. Determination of world plant formations from simple climatic data. Science. 105: 367–368.

Hole, F.D. 1978. An approach to landscape analysis with emphasis on soils. Geoderma. 21: 1–23.

Hole, F.D.; Campbell, J.B. 1985. Soil landscape analysis. Totowa, NJ: Rowman & Allanheld. 196p.

Hopkins, A.D. 1938. Bioclimatics: a science of life and climate relations. Misc. Publ. 280. Washington, DC: U.S. Department of Agriculture. 188p.

Hopkins, L.D. 1977. Methods for generating land suitability maps: a comparative evaluation. Journal American Institute of Planners. 43: 386–400.

Horton, R.E. 1967. Soil survey of Scotland County, North Carolina, USA. Washington, DC: USDA Soil Conservation Service. 70p. with 45 photomaps.

Host, G.E.; Pregitzer, K.S.; Ramm, C.W.; Hart, J.B.; Cleland, D.T. 1987. Landform-mediated differences in successional pathways among upland forest ecosystems in northwestern Lower Michigan. Forest Science. 33: 445–457.

Host, G.E.; Polzer, P.L.; Mladenoff, D.J.; White, M.A.; Crow, T.R. (1996) A quantitative approach to developing regional ecosystem classifications. Ecological Applications. 6: 608–818.

Howard, J.A.; Mitchell, C.W. 1985. Phytogeomorphology. New York: John Wiley. 222p.

Huang, S.; Prince, C.; Titus, S.J. 2000. Development of ecoregion-based height-diameter models for white spruce in boreal forests. Forest Ecology and Management. 129: 125–141.

Huggett, R.J. 1995. Geoecology: an evolutionary approach. London: Routledge. 320p.

Hume, T.M.; Snelder, T.; Weatherhead, M.; Liefting, R. 2007. A controlling factor approach to estuary classification. Ocean and Coastal Management. 50: 905–929.

Hunt, C.B. 1966. Plant ecology of Death Valley, California. Prof. Paper 509. Washington, DC: U.S. Geological Survey. 68p.

Hunt, C.B. 1967. Physiography of the United States. San Francisco: W.H. Freeman. 480p.

Hunt, C.B. 1972. Geology of soils: their evolution, classification, and uses. San Francisco: W.H. Freeman. 344p.

Hunt, C.B. 1974. Natural regions of the United States and Canada. San Francisco: W.H. Freeman. 725p.

Hustich, I. 1953. The boreal limits of conifers. Arctic. 6: 149–162.

Illies, J. 1974. Introduction to zoogeography (trans. from German by W.D. Williams). London: Macmillan Press. 120p.

Imbrie, J.; Imbrie, K.P. 1979. Ice ages: solving the mystery. Short Hills, NJ: Enslow Publishers. 224p.

IPCC. 2007. Climate change 2007: the physical science basis. Contribution of Working Group I to the Fourth Assessment Report of the Intergovernmental Panel on Climate Change (IPCC). Cambridge, UK: Cambridge University Press. 996p.

Isachenko, A.G. 1973. Principles of landscape science and physical-geographic regionalization (trans. from Russian by R.J. Zatorski, edited by J.S. Massey). Carlton, Victoria, Australia: Melbourne University Press. 311p.

Iverson, L.R.; Prasad, A.M. 2001. Potential changes in tree species richness and forest community types following climate change. Ecosystems. 4: 186–199.

Ives, J.D.; Messerli, B.; Spiess, E. 1997. Mountains of the world – A global priority. In: B. Messerli and J.D. Ives (eds.). Mountains of the world: a global priority. Pearl River, NY: Parthenon Publishing Group. pp. 1–15.

James, P.E. 1936. The geography of the oceans: a review of the work of Gerhard Schott. Geographical Review. 26: 664–669.

James, P.E. 1959. A geography of man. 2nd ed. Boston: Ginn. 656p.

Jepson, P.; Whittaker, R.J. 2002. Ecoregions in context: a critique with special reference to Indonesia. Conservation Biology. 16: 42–57.

Joerg, W.L.G. 1914. The subdivision of North America into natural regions: a preliminary inquiry. Annals Association of American Geographers. 4: 55–83.

Joerg, W.L.G. 1935. The natural regions of the world oceans according to Schott. Trans. American Geophysical Union, Sixteenth Annual Meeting, part I, April 25–26, 1935. Washington, DC: American Geophysical Union. pp. 239–245.

Johnson, K.N.; Stuart, T.W.; Crim, S.A. 1986. Forplan version 2: an overview. Washington, DC: USDA Forest Service, Land Management Planning Staff. Irregular pagination.

Kalesnik, S.V. 1962. Landscape science. In: C.D. Harris (ed.). Soviet geography, accomplishments and tasks. Am. Geog. Soc. Occasional Publ. No. 1. New York: American Geographical Society. pp. 201–204.

Kalvova, J.; Halenka, T.; Bezpalcova, K.; Nemesova, I. 2003. Köppen climate types in observed and simulated climates. Studia Geophysica et Geodaetica. 47: 185–202.

Kaufmann, M.R.; Graham, R.T.; Boyce, D.A.; Moir, W.H.; Perry, L.; Reynolds, R.T.; Bassett, R.L.; Mehlhop, P.; Edminster, C.B.; Block, W.M.; Corn, P.S. 1994. An ecological basis for ecosystem management. General Tech. Rep. RM-246. Fort Collins, CO: USDA Forest Service, Rocky Mountain Forest and Range Experiment Station. 22p.

Klijn, F.; Udo de Haes, H.A. 1994. A hierarchical approach to ecosystems and its applications for ecological land classification. Landscape Ecology. 9: 89–104.

Klopatek, J.M.; Olson, R.J.; Emerson, C.J.; Joness, J.L. 1979. Land-use conflicts with natural vegetation in the United States. Environmental Sciences Division Publ. No. 1333. Oak Ridge, TN: Oak Ridge National Laboratory. 19p.

Knight, D.H.; Reiners, W.A. 2000. Natural patterns in southern Rocky Mountain landscapes and their relevance to forest management. In: D.H. Knight, F.W. Smith, S.W. Buskirk, W.H. Romme, and W.L. Baker (eds.). Forest fragmentation in the Southern Rocky Mountains. Boulder, CO: University Press of Colorado. pp. 15–30.

Köppen, W. 1931. Grundriss der Klimakunde. Berlin: Walter de Gruyter. 388p.

Krajina, V.J. 1965. Biogeoclimatic zones and classification of British Columbia. In: V.J. Krajina (ed.). Ecology of western North America. Vancouver, British Columbia: University of British Columbia Press. pp. 1–17.

Kruckeberg, A.R. 2002. Geology and plant life: the effects of landforms and rock types on plants. Seattle, Washington: University of Washington Press. 362p.

Küchler, A.W. 1964. Potential natural vegetation of the conterminous United States. Spec. Publ. 36. New York: American Geographical Society. 116p. with separate map at 1:3,168,000.

Küchler, A.W. 1973. Problems in classifying and mapping vegetation for ecological regionalization. Ecology. 54: 512–523.

Küchler, A.W. 1974. Boundaries on vegetation maps. In: R. Tüxen (ed.). Tatsachen und Probleme der Grenzen in der Vegetation. Lehre, Germany: Verlag von J. Cramer. pp. 415–427.

Laut, P.; Paine, T.A. 1982. A step towards an objective procedure for land classification and mapping. Applied Geography. 2: 109–126.

Lightfoot, D.C.; Brantley, S.L.; Allen, C.D. 2008. Geographic patterns of ground-dwelling arthropods across an ecoregional transition in the North American Southwest. Western North American Naturalist. 68: 83–102.

Littell, J.S.; McKenzie, D.; Peterson, D.L.; Westering, A.l. 2009. Climate and wild-fire area burned in western U.S. ecoprovinces, 1916–2003. Ecological Applications. 14: 1003–1021.

Leser, H. 1976. Landschaftsökologie. Stuttgart: Eugen Ulmer. 432p.

Lewis, G.M. 1966. Regional ideas and reality in the Cis-Rocky Mountain west. Transactions of the Institute of British Geographers. 38: 135–150.

Lindsay, S.W.; Bayoh, M.N. 2004. Mapping members of the Anopheles gambiae complex using climate data. Physiological Entomology. 29: 204–209.

Lohmann, U.; Sausen, R.; Bengtsson, L.; Cubasch, U.; Perlwitz, J.; Roeckner, E. 1993. The Köppen climate classification as a diagnostic tool for general circulation models. Climate Research. 3: 177–193.

Lotspeich, F.B.; Platts, W.S. 1982. An integrated land-aquatic classification system. North American Journal of Fisheries Management. 2: 138–149.

Love, J.D.; Christiansen, A.C. 1985. Geologic map of Wyoming. Washington, DC: U.S. Geological Survey. 1:500,000; colored.

Loveland, T.R.; Merchant, J.W.; Ohlen, D.O.; Brown, J.F. 1991. Development of a land-cover characteristics database for the conterminous US. Photogrammetric Engineering and Remote Sensing. 57: 1453–1463.

Lowell, K.E. 1990. Differences between ecological land type maps produced using GIS or manual cartographic methods. Photogrammetric Engineering and Remote Sensing. 56: 169–173.

Lugo, A.E.; Brown, S.L.; Dodson, R.; Smith, T.S.; Shugart, H.H. 1999. The Holdridge life zones of the conterminous United States in relation to ecosystem mapping. Journal of Biogeography 26: 1025–1038.

Lugo, A.E.; Brown, S.L.; Dodson, R.; Smith, T.S.; Shugart, H.H. 2006. Long-term research at the USDA forest service's experimental forests and ranges. Bioscience. 56(1): 39–48.

MacArthur, R.H. 1972. Geographical ecology: patterns in the distribution of species. Princeton, NJ: Princeton University Press. 269p.

Major, J. 1951. A functional, factorial approach to plant ecology. Ecology. 32: 392–412.

Malamud, B.D.; Millington, J.D.A.; Perry, G.W. 2005. Characterizing wildfire regimes in the United States. Proceedings, National Academy of Sciences. 102: 4694–4699.

Malanson, G.P.; Butler, D.R. 2002. The Western Cordillera. In: Orme, A.R. (ed.). The physical geography of North America. New York: Oxford University Press. pp. 363–379.

Marschner, F.J. 1950. Major land uses in the United States. Washington, DC: USDA Bureau of Agricultural Economics. 1:5,000,000; colored.

Marston, R.A. 2006. President's column: Ecoregions: a geographic advantage in studying environmental change. Association of American Geographers Newsletter. 41(3): 3–4.

Mather, J.R.; Sdasyuk, G.V. (eds.). 1991. Global change: geographical approaches. Tucson, AZ: University of Arizona Press. 289p.

Matthews, E. 1983. Global vegetation and land use: new high-resolution data bases for climate studies. Journal of Climate and Applied Meterology. 22: 474–487.

McCreadie, J.W.; Adler, P.H. 2006. Ecoregions as predictors of lotic assemblages of blackflies (Diptera: Simuliidae). Ecography. 29: 603–613.

McHarg, I.L. 1969. Design with nature. Garden City, NY: American Museum of Natural History by The Natural History Press. 197p.

McMahon, G.; Gregonis, S.M.; Waltman, S.W.; Omernik, J.M.; et al. 2001. Developing a spatial framework of common ecological regions for the conterminous United States. Environmental Management. 28: 293–316.

McNab, W.H. 1989. Terrain shape index: quantifying effect of minor landforms on tree height. Forest Science. 35: 91–104.

McNab, W.H. 1991. Predicting forest type in Bent Creek Experimental Forest from topographic variables. In: S.S. Coleman and D.C. Neary (eds.). Proceedings of the sixth biennial southern silvicultural research conference. October 30–November 1, 1990. Memphis, Tennessee. General Tech. Rep. SE-70. Asheville, NC: Southeastern Forest Experiment Station. pp. 496–504.

McRoberts, R.E.; Nelson, M.D.; Wendt, D.G. 2002. Stratified estimation of forest area using satellite imagery, inventory data, and the k-Nearest Neighbors technique. Remote Sensing of Environment. 82: 457–468.

Meentemeyer, V.; Box, E.O. 1987. Scale effects in landscape studies. In: M.G. Turner (ed.). Landscape heterogeneity and disturbance. Ecological studies. vol. 64. New York: Springer-Verlag. pp. 15–34.

Merriam, C.H. 1890. Results of a biological survey of the San Francisco Mountain region and desert of the Little Colorado, Arizona. North American Fauna. 3: 1–136.

Merriam, C.H. 1898. Life zones and crop zones of the United States. Bull. Div. Biol. Surv. 10. Washington, DC: U.S. Department of Agriculture. pp. 1–79.

Milanova, E.V.; Kushlin, A.V. (eds.). 1993. World map of present-day landscapes: an explanatory note. Moscow: Moscow State University. 33p. with separate map at 1:15,000,000.

Mil'kov, F.N. 1979. The contrastivity principle in landscape geography. Soviet Geography. 20: 31–40.

Miller, A.A. 1946. Climatology. London: Methuen. 320p.

Miller, D.H. 1978. The factor of scale: ecosystem, landscape mosaic, and region. In: K.A. Hammond, G. Macinko, and W.B. Fairchild (eds.). Sourcebook on the environment. Chicago: University of Chicago Press. pp. 63–88.

Miller, J.R.; Turner, M.G.; Smithwich, E.A.H.; Dent, C.L.; Stanley, E.H. 2004. Spatial extrapolation: the science of predicting ecological patterns and processes. BioScience. 54(4): 310–320.

Milne, G. 1936. A provisional soil map of East Africa. East African Agric. Res. Sta., Amani Memoirs, 34p.

Mitchell, C.W. 1973. Terrain evaluation. London: Longman. 221p.

Mitchell, J.M. 1977. Carbon dioxide and future climate. Environmental Data Service, March. pp. 3–9.

Moss, M.R. 1985. Land processes and land classification. Journal of Environmental Management. 20: 295–319.

Muller, R.A.; Oberlander, T.M. 1978. Physical geography today: a portrait of a planet. 2nd ed. New York: Random House. 590p.

Mutel, C.F.; Emerick, J.C. 1992. From grassland to glacier. 2nd ed. Boulder, CO: Johnson Books. 290p.

The Nature Conservancy, 1997. Designing a geography of hope: guidelines for ecoregion-based conservation in The Nature Conservancy. Arlington, VA: The Nature Conservancy. 84p.

Neff, E. 1967. Die theoretischen grundlagen der landschaftslehre. Gotha. 152S.

Nielson, R.P. 1987. Biotic regionalization and climatic controls in western North America. Vegetatio. 70: 135–147.

Neilson, R.P. 1995. A model for predicting continental-scale vegetation distribution and water balance. Ecological Applications. 5: 362–385.

Nesser, J.A.; Ford, G.L.; Maynard, C.L.; Page-Dumroese, D.S. 1997. Ecological units of the Northern Region: subsections. Gen. Tech. Rep. INT-GTR-369. Ogden, UT: USDA Forest Service, Intermountain Research Station. 88p.

Noss, R.G.; LaRoe, E.T.; Scott, J.M. 1995. Endangered ecosystems of the United States: a preliminary assessment of loss and degradation. Biological Rep. 28. Washington, DC: National Biological Service. 58p.

O'Brien, R.A. 1996. Forest resources of northern Utah ecoregions. Resource Bulletin INT-RB-87. Ogden, UT: USDA Forest Service, Intermountain Research Station. 43p.

Odom, R.H.; McNab, W.H. 2000. Using digital terrain modeling to predict ecological types in the Balsam Mountain of western North Carolina. Research Note SRS-8. Asheville, NC: USDA Forest Service, Southern Research Station. 11p.

Odum, E.P. 1971. Fundamentals of ecology. 3rd ed. Philadelphia: W.B. Saunders. 574p.

Odum, E.P. 1977. The emergence of ecology as a new integrative discipline. Science. 195: 1289–1293.

Olson, C.G.; Hupp, C.R. 1986. Coincidence and spatial variability of geology, soils, and vegetation, Mill Run Watershed, Virgina. Earth Surface Processes and Landforms. 11: 619–629.

Olson, D.M.; Dinerstein, E.; Wikramanayake, E.D.; Burgess, N.D.; et al. 2001. Terrestrial ecoregions of the world: a new map of life on Earth. BioScience. 51(11): 933–938.

Olson, J.S.; Watts, J.S. 1982. Major world ecosystem complexes. In: Carbon in live vegetation of major world ecosystems. ORNL-5862. Oak Ridge, TN: Oak Ridge National Laboratory. 1:30,000,000.

Omernik, J.M. 1987. Ecoregions of the conterminous United States (map supplement). Annals Association of American Geographers. 77: 118–125.

Omernik, J.M. 2004. Perspectives on the nature and definition of ecological regions. Environmental Management. 34(S1): S27–S38.

Omernik, J.M.; Griffith, G.E. 1991. Ecological regions versus hydrologic units: frameworks for managing water quality. Journal of Soil and Water Conservation. 46: 334–340.

Omi, P.N.; Wensel, L.C.; Murphy, J.L. 1979. An application of multivariate statistics to land-use planning: classifying land units into homogeneous zones. Forest Science. 25: 399–414.

O'Neill, R.V.; DeAngelis, D.L.; Waide, J.B.; Allen, T.F.H. 1986. A hierarchical concept of ecosystems. New Jersey: Princeton University Press. 253p.

Oosting, H.J. 1956. The study of plant communities. 2nd ed. San Francisco: W.H. Freeman. 440p.

Orme, A.T.; Bailey, R.G. 1971. Vegetation and channel geometry in Monroe Canyon, southern California. Yearbook of the Association Pacific Coast Geographers. 33: 65–82.

Passarge, S. 1929. Die Landschaftsgürtel der Erde, Natur und Kultur. Breslau: Ferdinand Hirt. 144p.

Peet, R.K. 1981. Forest vegetation of the Colorado Front Range. Vegetatio. 45: 3–75.

Peet, R.K. 1978. Latitudinal variation in southern Rocky mountain forests. Journal of Biogeography. 5: 275–289.

Peet, R.K. 1988. Forests of the Rocky Mountains. In: M.G. Barbour and W.D. Billings (eds.). North American terrestrial vegetation. Cambridge, England: Cambridge University Press. pp. 63–102.

Pfister, R.D.; Arno, S.F. 1980. Classifying forest habitat types based on potential climax vegetation. Forest Science. 26: 52–70.

Pflieger, W.L. 1971. A distributional study of Missouri fishes. University of Kansas Publication Museum of Natural History 20: 225–570.

Pojar, J.; Klinka, K.; Meidinger, D.V. 1987. Biogeoclimatic ecosystem classification in British Columbia. Forest Ecology and Management. 22: 119–154.

Polunin, N.; Worthington, E.B. 1990. On the use and misuse of the term "ecosystem." Environmental Conservation. 17: 274.

Powell, D.S.; Faulkner, J.L.; Darr, D.R.; Zhu, Z.; MacCleery, D.W. 1993. Forest resources of the United States, 1992. General Tech. Rep. RM-234. Fort Collins, CO: Rocky Mountain Forest and Range Experiment Station. 132p. with separate map at 1:7,500,000.

Pu, R.; Li, Z.; Gong, P.; Csiszar, I.; Fraser, R.; et al. 2007. Development and analysis of a 12-year daily 1-km forest fire dataset across North America from NOAA/AVHRR data. Remote Sensing of Environment. 108: 198–208.

Prentice, I.C.; Cramer, W.; Harrison, S.P.; Leemans, R.; Monserud, R.A.; Solomon, A.M. 1992. A global biome model based on plant physiology and dominance, soil properties and climate. Journal of Biogeography. 19: 117–134.

Rabeni, C.F.; Doisy, K.E. 2000. Correspondence of stream bentic invertebrate assemblages to regional classification schemes in Missouri. Journal of the North American Benthological Society. 19: 419–428.

Radloff, D.L.; Betters, D.R. 1978. Multivariate analysis of physical site data for wildland classification. Forest Science. 24: 2–10.

Risser, P.G. 1993. Ecotones at local to regional scales from around the world. Ecological Applications. 3: 367–368.

Robertson, J.K.; Wilson, J.W. 1985. Design of the national trends network for monitoring the chemistry of atmospheric precipitation. Circular 964. Washington, DC: U.S. Geological Survey. 46p.

Robinove, C.J. 1979. Integrated terrain mapping with digital Landsat images in Queensland, Australia. Prof. Paper 1102. Washington, DC: U.S. Geological Survey. 39p.

Rowe, J.S. 1961. The level-of-integration concept and ecology. Ecology. 42: 420–427.

Rowe, J.S. 1979. Revised working paper on methodology/philosophy of ecological land classification in Canada. In: C.D.A. Rubec (ed.). Applications of ecological (biophysical) land classification in Canada. Ecological Land Classification Series No. 7. Ottawa: Environment Canada. pp. 23–30.

Rowe, J.S. 1980. The common denominator in land classification in Canada: an ecological approach to mapping. Forestry Chronicle. 56: 19–20.

Rowe, J.S.; Sheard, J.W. 1981. Ecological land classification: a survey approach. Environmental Management. 5: 451–464.

Rowe, J.S. 1996. Land classification and ecosystem classification. Environmental Monitoring and Assessment. 39: 11–20.

Rudis, V.A. 1998. Regional forest resource assessment in an ecological framework: The Southern United States. Natural Areas Journal. 18: 319–332.

Rudis, V.A. 1999. Ecological subregion codes by county, coterminous United States. General Technical Report SRS-36. Asheville, NC:USDA Forest Service, Southern Research Station. 95p.

Ruhe, R.V. 1960. Elements of the soil landscape. 7th International Congress of Soil Science. 23: 165–170.

Salwasser, H. 1990. Conserving biological diversity: a perspective on scope and approach. Forest Ecology and Management. 35: 79–90.

Sauer, C.O. 1925. The morphology of landscape. University of California Publications in Geography. 2: 19–53.

Schmithüsen, J. 1976. Atlas zur Biogeographie. Mannheim-Wien-Zurich: Bibliographisches Institut. 88p.

Schultz, A.M. 1967. The ecosystem as a conceptual tool in the management of natural resources. In: S.V.C. Wantrup and J.S. Parsons (eds.). Natural resources: quality and quantity. Berkeley: University of California Press. pp. 139–161.

Schultz, J. 1995. The ecozones of the World: the ecological divisions of the geosphere (trans. from German by I. and D. Jordan). Berlin: Springer-Verlag. 449p.

Schwartz, M.K.; Mukksm, L.S.; Ortega, Y.; Ruggiero, L.F.; Allendorf, F.W. 2003. Landscape location affects genetic variation of Canada lynx (Lynx canadensis). Molecular Ecology. 12: 1807–1816.

Schott, G. 1936. Die Aufteilung der drei Ozeane in natürliche Regionen. Petermann's Mitteilungen. 82: 165–170; 218–222.

Sellers, W.D. 1965. Physical climatology. Chicago: The University of Chicago Press. 272p.

Silbernagel, J. 2005. Bio-regional patterns and spatial narratives for integrative landscape research and design.In: B. Tress, G. Tres, G. Fry, and P. Opdam (eds.). From Landscape research to landscape planning: aspects of integration, education, and application. Wageningen UR Frontis Series, vol. 12. Dordrecht: Springer. pp. 107–118.

Smith, R.L. 1977. Elements of ecology and field biology. New York: Harper & Row. 497p.

Snelder, T.H.; Biggs, B.J.F.; Woods, R.A. 2005. Improved eco-hydrological classification of rivers. River Research and Applications. 21: 609–628.

Sokal, R.R. 1974. Classification: purposes, principles, progress, prospects. Science. 185: 1115–1123.

Soriano, A.; Paruelo, J.M. 1992. Biozones: vegetation units defined by functional characters identifiable with the aid of satellite sensor images. Global Ecology and Biogeography Letters. 2: 82–89.

Spalding, M.D.; Fox, H.E.; Allen, G.R.; Davidson, N. 2007. Marine ecoregions of the world: a bioregionalization of coastal and shelf areas. Bioscience. 57: 573–583.

Stephenson, N.L. 1990. Climatic control of vegetation distribution: the role of the water balance. American Naturalist. 135: 649–670.

Strahler, A.N. 1965. Introduction to physical geography. New York: John Wiley. 455p.

Strahler, A.N.; Strahler, A.H. 1976. Elements of physical geography. New York: John Wiley. 469p.

Strahler, A.H.; Strahler, A.N. 1996. Introducing physical geography, environmental update. New York: John Wiley. 565p.

Sukachev, V.; Dylis, N. 1964. Fundamentals of forest biogeocoenology (trans. from Russian by J.M. Maclennan). London: Oliver & Bond. 672p.

Swan, L.W. 1967. Alpine and aeolian regions of the world. In: H.E. Wright, Jr., and W.H. Osburn (eds.). Arctic and alpine environments. Bloomington, IN: Indiana University Press. pp. 29–54.

Swanson, F.J.; Kratz, T.K.; Caine, N.; Woodmansee, R.G. 1988. Landform effects on ecosystem patterns and processes. Bioscience. 38: 92–98.

Swanson, F.J.; Franklin, J.F.; Sedell, J.R. 1991. Landscape patterns, disturbance, and management in the Pacific Northwest, USA. In: I.S. Zonneveld and R.T.T. Forman (eds.). Changing landscapes: an ecological perspective. New York: Springer-Verlag. pp. 191–213.

Swift, M.J.; Heal, O.W.; Anderson, J.M. 1979. Decomposition in terrestrial ecosystems. Studies in Ecology, vol. 5. Oxford: Blackwell Scientific Publication. 384p.

Szekely, A. 1992. Establishing a region for ecological cooperation in North America. Natural Resources Journal. 32: 563–622.

Tansley, A.G. 1935. The use and misuse of vegetation terms and concepts. Ecology. 16: 284–307.

Thayer, R.L. 2003. LifePlace: bioregional thought and practice. Berkeley: University of California Press. 300p.

Thornthwaite, C.W. 1931. The climates of North America according to a new classification. Geographical Review. 21: 633–655, with separate map at 1:20,000,000.

Thornthwaite, C.W. 1948. An approach toward a rational classification of climate. Geographical Review. 38: 55–94.

Thornthwaite, C.W. 1954. Topoclimatology. In: Proceedings of the Toronto meteorological conference, September 9–15, 1953. Toronto: Royal Meteorological Society. pp. 227–232.

Thorp, J. 1931. The effects of vegetation and climate upon soil profiles in northern and northeastern Wyoming. Soil Science. 32: 290.

Thrower, N.J.W.; Bradbury, D.E. 1973. The physiography of the Mediterranean landscape with special emphasis on California and Chile. In: F. di Castri and H.A. Mooney (eds.). Mediterranean type ecosystems: origin and structure. Ecological studies: analysis and synthesis. vol. 7. Heidelberg: Springer-Verlag. pp. 37–52.

Tosi, J.S. 1964. Climatic control of terrestrial ecosystems: a report on the Holdridge model. Economic Geography. 40: 173–181.

Trewartha, G.T. 1968. An introduction to climate. 4th ed. New York: McGraw-Hill. 408p.

Trewartha, G.T.; Robinson, A.H.; Hammond, E.H. 1967. Physical elements of geography. 5th ed. New York: McGraw-Hill. 527p.

Tricart, J.; Cailleux, A. 1972. Introduction to climatic geomorphology (trans. from French by C.J. Kiewiet de Jonge). New York: St. Martin's Press. 274p.

Tricart, J.; Kiewiet de Jonge, C. 1992. Ecogeography and rural management. Essex, England: Longman. 267p.

Troll, C. 1964. Karte der Jahrzeiten-Klimate der Erde. Erdkunde. 17: 5–28.

Troll, C. 1966. Seasonal climates of the earth. The seasonal course of natural phenomena in the different climatic zones of the earth. In: E. Rodenwaldt and H.J. Jusatz (eds.). World maps of climatology. 3rd ed. Berlin: Springer-Verlag. pp. 19–28. with separate map at 1:45,000,000 by C. Troll and K.H. Paffen.

Troll, C. 1968. The Cordilleras of the tropical Americas, aspects of climatic, phytogeographical and agrarian ecology. In: C. Troll (ed.). Geo-ecology of the mountainous regions of the tropical Americas. Bonn, Ferd. Dümmlers Verlag. pp. 15–56.

Troll, C. 1971. Landscape ecology (geoecology) and biogeocenology—a terminology study. Geoforum. 8: 43–46.

Troll, C. 1972. Geoecology and the world-wide differentiation of high-mountain ecosystems. In: Troll, C. (ed.) Geoecology of the high-mountain regions of Eurasia. Wiesbaden: Franz Steiner Verlag. pp. 1–13.

Tucker, C.J.; Townshend, J.R.G.; Goff, T.E. 1985. African land-cover classification using satellite data. Science. 227: 369–375.

Turner, M.G.; Gardner, R.H.; O'Neill, R.V. 2001. Landscape ecology in theory and practice. New York: Springer-Verlag. 401p.

Udvardy, M.D.F. 1975. A classification of the biogeographical provinces of the world. Occasional Paper No. 18. Morges, Switzerland: International Union for Conservation of Nature and Natural Resources. 48p.

USDA Forest Service. 1982. Ecosystem classification, interpretation, and application. Forest Service Manual 2061, 2062, 2063. Washington, DC.

USDA Soil Conservation Service. 1975. Soil taxonomy: a basic system for making and interpreting soil surveys. Agric. Handbook 436. Washington, DC: U.S. Department of Agriculture. 754p.

U.S. General Accounting Office. 1994. Ecosystem management: additional actions needed to adequately test a promising approach. GAO/RCED-94–111 Ecosystem Management. Washington, DC: U.S. General Accounting Office. 87p.

U.S. Geological Survey. 1970. National atlas of the United States of America, Washington, DC.

U.S. Geological Survey. 1979. Accounting units of the national water data network, Washington, DC. 1:7,500,000.

Vale, T.R. 1982. Plants and people: vegetation change in North America. Washington, DC: Association of American Geographers. 88p.

van der Maarel, E. 1976. On the establishment of plant community boundaries. Bericht der Deutschen botanischen Gesellschaft. 89: 415–433.

Van der Ryn, S.; Cowan, S. 1996. Ecological design. Washington, DC: Island Press. 201p.

Van Dyne, G.M. (ed.). 1969. The ecosystem concept in natural resource management. New York: Academic Press. 383p.

Vaughan, T.A. 1954. Mammals of the San Gabriel Mountains of California. University of Kansas Publications, Museum of Natural History. 7: 513–582.

Vaughan, T.A.; Ryan, J.M.; Czaplewski, N.J. 2000. Mammalogy. 4th ed. Pacific Grove, CA: Brooks/Cole. 565p.

Veatch, J.O. 1930. Natural geographic divisions of land. Michigan Academy of Sciences, Arts and Letters. 19: 417–427.

Viereck, L.A.; Dyrness, C.T.; Batten, A.R.; Wenzlick, K.L. 1992. The Alaska vegetation classification. General Tech. Rep. PNW-GTR-286. Portland, OR: USDA Forest Service, Pacific Northwest Research Station. 278p.

Vogel, K.P.; Schmer, M.R.; Mitchell, R.B. 2005. Plant adaptation regions: ecological and climatic classification of plant materials. Rangeland Ecology & Management. 58: 351–319.

Walker, D.A. 2000. Hierarchical subdivision of Arctic tundra based on vegetation response to climate, parent material and topography. Global Change Biology. 6(1): 19–34.

Walter, H. 1977. Vegetationszonen und Klima: die ökologische Gliederung der Biogeosphäre. Stuttgart, Germany: Eugen Ulmer Verlag. 309p.

Walter, H. 1984. Vegetation of the earth and ecological systems of the geobiosphere (trans. from German by O. Muise). 3rd ed. Berlin: Springer-Verlag. 318p.

Walter, H.; Box, E. 1976. Global classification of natural terrestrial ecosystems. Vegetatio. 32: 75–81.

Walter, H.; Breckle, S.-W. 1985. Ecological Systems of the geobiosphere, vol. 1, Ecological principles in global perspective (trans. from German by S. Gruber). Berlin: Springer-Verlag. 242p.

Walter, H.; Harnickell, E.; Mueller-Dombois, D. 1975. Climate-diagram maps of the individual continents and the ecological climatic regions of the earth. Berlin: Springer-Verlag. 36p. with 9 maps.

Walter, H.; Lieth, H. 1960–1967. Klimadiagramm weltatlas. Jena, East Germany: G. Fischer Verlag. Maps, diagrams, profiles. Irregular pagination.

Wang, M.; Overland, J.E. 2004. Detecting arctic climate change using Köppen climate classification. Climatic Change. 67: 43–62.

Warren, C.E. 1979. Toward classification and rationale for watershed management and stream protection. EPA-600/3–79–059. Corvallis, OR: U.S. Environmental Protection Agency. 142p.

Webster, J.S. 1979. Hierarchical organization of ecosystems. In: E. Halfon (ed.). Theoretical systems ecology. New York: Academic Press. pp. 119–129.

Wertz, W.A.; Arnold, J.F. 1972. Land systems inventory. Ogden, UT: USDA Forest Service, Intermountain Region. 12p.

Westman, W.E. 1985. Ecology, impact assessment, and environmental planning. New York: John Wiley. 532p.

Whittaker, R.H. 1975. Communities and ecosystems. 2nd ed. New York: MacMillan. 387p.

Whittaker, R.H.; Niering, W.A. 1965. Vegetation of the Santa Catalina Mountains, Arizona: a gradient analysis of the south slope. Ecology. 46: 429–452.

Wiens, J.A.; Crawford, C.S.; Gosz, J.R. 1985. Boundary dynamics: a conceptual framework for studying landscape ecosystems. Oikos. 45: 421–427.

Wiken, E.B.; Ironside, G. 1977. The development of ecological (biophysical) land classification in Canada. Landscape Planning. 4: 273–275.

Wolfe, J.N.; Wareham, R.T.; Scofield, H.T. 1949. Microclimates and macroclimate of Neotoma, a small valley in central Ohio. Bulletin Ohio Biological Survey. 8: 1–267.

Woodall, C.W.; Liknes, G.C. 2008. Climatic regions as an indicator of forest coarse and fine woody debris carbon stocks in the United States. Carbon Balance and Management 3:5. Available at http://www.cbmjournal.com/content/3/1/5

Woodward, J. 2000 Waterstained landscapes: seeing and shaping regionally distinctive places. Baltimore, MD: John Hopkins University Press. 221p.

World Conservation Monitoring Centre. 1992. Global biodiversity: status of the Earth's living resources. London: Chapman & Hall. 594p.

Yoshino, M.M. 1975. Climate in a small area: an introduction to local meteorology. Tokyo: University of Tokyo Press. 549p.

Zonneveld, I.S. 1972. Land evaluation and land(scape) science. Use of aerial photographs in geography and geomorphology. ITC textbook of photointerpretation. vol. VII. Enschede, The Netherlands: International Training Centre for Aerial Survey. 106p.

# About the Author

**Robert G. Bailey** (b. 1939) received his PhD in Geography from the University of California, Los Angeles (1971). A geographer with the U.S. Forest Service, Rocky Mountain Research Station, he was leader of the agency's Ecosystem Management Analysis Center for many years. He has four decades of experience working with the theory and practice of ecosystem classification and mapping, with applications in slope stability, land capability, inventory and monitoring, ecosystem management, climate change, and sustainability. He is author of numerous publications on this and related subjects, including three books.

# Subject Index

Printed in the United States of America

MAR 1 2011